T0129653

Springer Praxis Books

Popular Science

More information about this series at http://www.springer.com/series/8158

Peter Hulsroj

What If We Don't Die?

The Morality of Immortality

Copernicus Books is a brand of Springer

Published in association with
Praxis Publishing
Chichester, UK

Peter Hulsroj
Vienna, Austria

Springer Praxis Books
ISBN 978-3-319-19092-1 ISBN 978-3-319-19093-8 (eBook)
DOI 10.1007/978-3-319-19093-8

Library of Congress Control Number: 2015949996

Springer Cham Heidelberg New York Dordrecht London

Cover illustration: Joris Thys

Printed on acid-free paper

Copernicus Books is a brand of Springer
Springer International Publishing AG Switzerland is part of Springer Science+Business Media (www.springer.com)

Foreword

In the beginning was the Word, and the Word was with God, and the Word was God.

Hearing this every author must take heart. The only tool of the profession is primordial and divine. 'In the beginning was the Word' should be the tattoo on every author's forearm.

Of course, even if the word was the beginning, it does not mean that words are eternal. Nevertheless, in an ephemeral world surely many who like to write are comforted by the thought that what is written is something durable. A love letter may surface long after love has died. Yet, the lost emotion is glued to the words, be they simple or sophisticated.

Every book has an aspiration, and the aspiration is carried by words. They may not be immortal, but they are instigators, comforters, unsettlers, dumb, wise, sweet, sour, dreamy or crass. The words are not immortal, but as they provoke emotion, reasoning and action, they lead to something immortal, because every emotion and every action that result will reverberate forever, dimmer and dimmer with time that is true, but still forever. Each word has its own butterfly effect!

The book I have written seeks to hold the mirror of immortality to our value systems. By doing so rich colours come out, rather than the black and white of death (bad) and immortality (good). The book holds no claim to immortality for itself, but hopes to have set free a many-splendoured butterfly!

The responsibility for infelicities and errors in this book is mine and mine alone. However, should you find the book entertaining or of some utility, much of the credit goes to my friend and colleague Marco Aliberti. When I had done the first draft, I asked Marco to review it, which he did enthusiastically. And back he came with new ideas, new sources, lots and lots of learning and wisdom. It is true to say that often I turned his inspiration into something quite different from what he intended, which he accepted gracefully, but without Marco's prompting, so many interesting ideas and issues would have been missed.

The fantastic feature of family is how it earths and at the same time gives wings! I am immensely grateful for the fulfilment my wife and daughters have given me. I learn from their common sense, their hunger for knowledge and their boundless imagination, and I hope this book is a decent reflection of what I have learned from them; from my parents, so fearless and restless in their inquiry; and from so many others. No man is an island – and I least of all!

Vienna
July 2nd, 2015

Peter Hulsroj

Contents

Chapter 1
Immortality Again?

Is there any topic more alive than death – any desire more vivid than our hope to defy the grim reaper? Hopes and dreams of immortality permeate our lives. The mind-time we devote to immortality and to its close relative, eternity, is less than the attention we shower on money, sex and love, but is hardly surpassed by any other human topic. And, in a round-about fashion, we address immortality and eternity also when we think about money, sex and love. Money, sex and love might appear to immerse us in immediacy, and yet the immediate is often an escape from thinking about whether we are mortal and temporal, or, as we hope, immortal and eternal. The chasing of money and sex, in particular, is found by many to be the most probate means of avoiding the dark thoughts of the beyond and the 'fear and trembling'[1] involved; Mozart illustrates the mind-set superbly in Don Giovanni with the aesthete, immersed in immediacy, staying true to his creed even in the face of the Commander and death itself.

In human conversations and labelling the words *immortal* and *eternal* crop up incessantly, sometimes in the most interesting, unusual and amusing places. We might be gripped by the Lux Aeterna[2] of Requiem Masses, impressed by the 'eternal city' or 'immortal love', entertained by 'the immortal Beatles song, Yesterday', but be surprised about Eterna watches (even in eternity we must keep time!) and Eterna shirts, while being concerned about having to pay 'perpetuities'.

[1] See the eponymous book by Soeren Kierkegaard.

[2] The idea of light, the most ephemeral of phenomena, as a messenger of the eternal is full of tension, yet it is beautifully fitting that the immediacy of the moment should be the harbinger of an immediacy of the eternal.

P. Hulsroj, *What If We Don't Die?*, Springer Praxis Books,
DOI 10.1007/978-3-319-19093-8_1

Lux Aeterna
(Hieronymus Bosch, Ascent to Heaven)

The Abrahamic religions carry with them the promise of eternal life,[3] the Greeks had the Hades of the shadows and popular youth literature abounds with stories of vampires and the undead. The Egyptian elite elegantly fashioned a religion where the pharaohs and their circle could gain immortality but the plebs would be left to rot both literally and religiously. Vikings did not want to cross the dark river on their own, bringing slaves, horses and treasure with them. The Chinese emperor Qin was escorted to the beyond by his terra cotta army, after he had lost his long quest to find the elixir of immortality. New Age philosophy finds death uncool but is highly uncommitted to its alternatives. Hindu and Buddhist religions have rebirth as a central tenet.

In sum, there is no shortage of thinking about the eternal and about immortality and the fabric of society and of our individual lives is interwoven with concepts of immortality and the language of the eternal. Is there anything new that can possibly be said?

In fact, we are much lacking in giving concrete shape to what we believe immortality would bring human beings. Our attraction to immortality is much more motivated by our rejection of death than by our desire for eternal life. And because we have no idea how eternal life would be, except that it would not include death, in reality we do not know whether it is desirable. This is a tragic irony. We spend so much time worrying about being temporal that we entirely forget to think about whether we would truly want to be eternal. Over the millennia millions have perished in religious wars and persecutions in order to not ultimately perish, thus facing their fears in the ultimate fashion, without being clear on what the hereafter is and whether it is desirable. Hinduism and Buddhism are, in most respects, clear on what awaits, Christianity most certainly is not.

The purpose of this book is, however, not so much to set out the topography of Paradise, as it is to investigate whether 'progress' will allow us to create a Paradise on Earth in which humans are immortal. And to investigate whether such an earthly Paradise will be desirable, sustainable and moral!

Investigations of this kind are few in our day and age. The reason for this is presumably that for a century and a half Western society has by and large been guided by technology and the physical sciences. This has brought us material prosperity and materialism and a focus on the provable. It has commoditised social values, aided by both communism and capitalism, vulgarised sex and flattened our intellectual lives more than we care to think about. The '68 rebellion against some of these values spectacularly failed to give birth to more than the coolest music. New Age ideology remains a US West Coast phenomenon, and the hippie legacy is solely a reluctance to accept authority, be it intellectual or otherwise. This is not bad, but far from a battle cry for spiritualism and true spiritual investigation. Religious investigation has become more cross-cultural, but it is telling that thoughts on the God gene come from the atheist camp, which, in keeping with its roots in the physical sciences, such as biology, has lots of wind in its sails, and nowadays seems

[3]And there is more than a little tension in the idea that immortality is achieved through the portal of mortality. In German the tension becomes acute, durch Sterben zu Unsterblichkeit (through death to deathlessness)!

to be the only forceful voice in Western religion discourse. The believers continue to be kept back by the misapprehension of 'blessed be those who have not seen and yet have believed', and the terrible logic of believing to have to reach a certain result in order not to be condemned as heretics. Despite the fraying of religious authority in the First World, or perhaps exactly because of the fraying, organised religion has become intellectually impoverished. The Catholic Church tends to rage against the sin of materialism, rather than explaining the beauty of the virtue of the spiritual. A religion that uses the material, think St. Peter's and the splendour of thousands of churches, as a stepping-stone to the spiritual clearly has difficulty in preaching anti-materialism.

When our society approaches issues such as eternity it follows that it does so with a mind-set heavily influenced by the physical sciences. The Big Bang might be discussed as might the Big Crunch, but we do not discuss in any meaningful philosophical way how 'nothing comes from nothing', as the song goes,[4] or how something that exists can disappear rather than just be transformed, although this is the fundamental credo of the physical sciences. We seem to believe that ultimately an Aristotelian scientific approach will give us the answers step by step. We believe that when we launch space-probes like Herschel and Planck that look back almost to the start of the Big Bang we will then be able to understand the fundamental issue of something out of nothing, but of course, we will not. We will find that the Universe is ever more incredible, but the physical sciences will hardly let us transcend the dimensions that shackle our understanding of beginning and end. Singularity will in all likelihood be a wall we cannot climb, unless metaphysical discovery will give us the means.

Society and its philosophers do not want to go out on a limb, do not dare to attempt to give answers without resort to provability, logical positivism, and because of this timidity the dreamers, which is what philosophers should be, will not help physical science, will not pick up the gauntlets that advances in the physical sciences continually throw down. How often did you hear philosophers, poets, novelists address the mysteries of the likely multi-verse, or ethicists the implications of the probability of extra-terrestrial life? And even less so the professors of religion and psychology.

Despite the multitude of deserving topics currently being ignored, this book has carved out only a small niche. Yet this small niche presents in itself a multitude of issues, some metaphysical, some practical and many emotional. Because of the contact points with many intellectual disciplines and many religious, moral and personal topics this book comes far afield, yet tries not to stray too far from its central core, which is how we should position ourselves relative to a future in which humans might be able to create their own immortality.

If you believe that this assumption is too outlandish you should not immediately bin the book, since the assumption might still give birth to interesting insights on a number of existential issues, even if you believe that the boundary conditions of existence will not change as radically as all that!

---

[4]The Sound of Music.

Chapter 2
Is That Really a Question??

A first principle of epistemology should be that reality is almost always wilder and more complex than we imagine. Reality offers far more possibility for both creation and destruction than we believe. Humankind moves from simple to more sophisticated truths. No, the sun does not circle the Earth. No, quantum mechanics is probably not the ultimate puzzle, and it is doubtful that the smallest particle has yet been found. We are scratching the surface of the enigmas of dark energy and dark matter only, and in the immaterial field we ponder soul in the same fashion as we ponder gravity – by observation rather than understanding. The legacy of Aristotle and David Hume!

'There are more things in heaven and earth, Horatio, Than are dreamt of in our philosophy'[1] was how Shakespeare so succinctly described the limits on our imagination 400 years ago. With our increasing knowledge this wisdom has become truer, rather than the opposite! 'And therefore as a stranger give it welcome'.

If we accept the suggested first principle of epistemology we must assume a multiverse rather than a universe, we must assume intelligent life beyond Earth, and we must assume leaps and bounds in the life-sciences, leading to the possibility of eternal life in the not too distant future. So, no, death might not be the ultimate reality.

In this book not much will be said about how immortality will come about. For that the mooted first principle of epistemology will be left to reign. Suffice it to say that even with our limited predictive skills revolutions in genetics and biotechnology make the possibility of immortality palpable. We will be likely to truly understand the drivers of ageing; we will be likely to be able to change the genetic makeup or to refresh our genetic pool; we will be able to prove Schopenhauer wrong when he said: The life of a man is a struggle for existence with the certainty of defeat.

Instead of looking at how immortality can be achieved through genetics and biology, this book will look at what immortality means in a more fundamental sense, whether immortality on Earth makes religion irrelevant, whether physical immortality will

[1] Hamlet Act 1, scene 5, 159–167.

P. Hulsroj, *What If We Don't Die?*, Springer Praxis Books,
DOI 10.1007/978-3-319-19093-8_2

permit spiritual immortality, and how immortality would affect all the parameters that have determined our existence so far, including our hard-coded interest in passing on our genes. By reflecting on this, the ultimate question becomes whether eternal life on Earth is truly desirable, whether it is moral and, perhaps, how eternal life on Earth would be different from eternal life in some beyond. And we come full circle to the question of what immortality is: is it the continued existence of some spiritual substance or the continued existence of consciousness. If we say continued consciousness does it follow that the fear of death of the demented is delusional, as consciousness is hardly continuous, or that soul does not reside in early childhood, as neither memory nor forward planning have developed? The terrible logic of Peter Singer in 'Unsanctifying human life'![2] If we say that the point is the continued existence of some spiritual substance, similar to the physical properties that make up our bodies, we face the question of whether death is then relevant at all. Assuming that nothing perishes, only changes form, which is the absolute credo of the natural sciences, why fear death? Then the person is part of a much wider fabric, at one point in one form, at another in a different form. This is the logic and dilemma of rebirth.

Shirley MacLaine might find comfort in her belief in having been many previous persons (all fabulous!), but this is in the normal case only a comfort if it will be remembered. Not many assume the possession of MacLaine's alleged death-transcending recollection skills. And even if you did remember being a water buffalo what use is that? If you could lift the curtain of successive existences, and truly remember, it would be the voyeur's satisfaction. The dove would remember the water buffalo fondly but it would not by any stretch of the imagination share any significant part of the water buffalo's existence or personality even if they were different expressions of the same spiritual matter. A good novel lets us share in the life of the characters, but it does not turn us into the characters. Neither would the novel of successive lives, even if we were always reborn as humans!

The logical challenge of the Hindu/Buddhist reincarnation 'fear' of having to go through a long string of lives until desire is ultimately extinguished is that the fear presupposes that there is a personal identifier which is independent of physical or psychological manifestation, and which must suffer through rebirth until the ultimate release from desire. But what do you care about the suffering of a future you, if in that future you share no consciousness and no physical property with the *you* of the earlier life? The vase that breaks might provide the material for yet another vase which might yet again break. But this does not mean that the first vase breaks twice. In Hindu/Buddhist thinking there is a sub-textual identifier and connector between lives, elaborated differently in the various philosophical schools, but memory or continued consciousness is not key, except perhaps at a moment of ultimate enlightenment.[3] Hence you as a substrate might be refined though a myriad of lives, but you as a *you* of continuously switched-on consciousness and memory

[2] The moral philosopher Peter Singer argues that individual human life should be protectable only at the point in time when is has developed rationality, autonomy and self-consciousness. As newborns do not possess these capabilities their lives should not be protected.

[3] See John Hicks, Death & Eternal Life, on the range of identifiers in Asian thought.

will not be carried forward. It could thus be argued that at the level of substrate fear might be justified (if this is at all imaginable), but at the level of the conscious human being it is not.

An immortality constituted by some form of retention of both physical and spiritual form on Earth still leaves many questions unanswered as to its actual nature, as shall be addressed later. But this kind of immortality also gives new life to Pascal's Wager, albeit in a perhaps surprising form. Pascal argued that a virtuous life full of deprivation was worth it given that the possible ultimate reward would be the unspeakable good of eternal life. And even assuming the considerable risk of God not existing and hence of immortality being unattainable, the deprivation is worth it, because risk is, by far, outweighed by possible reward. Assuming now that earthly immortality becomes an option it can be asked whether it will it always be best to remain Earth-bound rather than taking the Kierkegaardian leap of faith, in this case a leap to death, in order to achieve the possibly much greater happiness of living in God's abode?[4] A utilitarian would perhaps argue that the logical route would be to live as much of earthly life as is possible without being bored to death, and then, when bored to death, die – hoping or not for a further life with God. Yet, even looking at it in this way does not entirely resolve the dilemma, because the point in time when you would consider yourself bored to death will to some extent depend on whether you will assume, and hope for, a further life in the heavens. But also there would be a religious question on whether it could be assumed that God would keep the option of heavenly life open, as it were, forever:

> Now this I say, brethren, that flesh and blood cannot inherit the kingdom of God; neither doth corruption inherit incorruption.
>
> Behold, I shew you a mystery; We shall not all sleep, but we shall all be changed,
>
> In a moment, in the twinkling of an eye, at the last trump: for the trumpet shall sound, and the dead shall be raised incorruptible, and we shall be changed.[5]

What Paul seems to say is that the mortal cannot be turned immortal other than through the intervention of God, and, by implication, that God can also deny immortality; God can let the trumpet remain silent!

Some religious thought might suggest that when you make yourself God by seizing the power of life and death, then you remove yourself from God's remit: you become a rival God of sorts. Pascal's Wager in this form becomes ultra-existential in that God's demand might become a choice between certain earthly immortality and possible heavenly immortality, with one being tangible and understandable and one being sublimation.

---

[4] Victor Hugo said: 'When grace is joined with wrinkles, it is adorable. There is an unspeakable dawn in happy old age' – perhaps a poetic description of Kingdom come, that would be denied by earthly immortality?

[5] 1 Corinthians 15:22, 15:23.

Chapter 3
Choices, Choices

The topic of this book is not just immortality, but the morality thereof. Morality assumes choice, and hence fundamental questions arise, such as for whom the choice: society or the individual?

If we analyse the moral question facing society first, many will argue that there is no question. We know from the Bible that once we have bitten the apple of knowledge there is no way back. Curiosity and innovation are our most enduring characteristic as a species, so what we can achieve in terms of arguable progress we will ultimately also achieve. We assume that the wheels of progress must churn forevermore. If earthly immortality is within our reach then as a society we will grab immortality for those who want it.

The fact that this logic is depressing does not make it less compelling. The logic applied to immortality is a scaled-up version of the logic many assume to apply to other progress such as genetic manipulation be it with humans, animals or plants. Yet, one of the most waxing issues hurtling our way is whether what used to be understood as progress is still to be understood as progress. This is a dialectic that started in earnest with the uncorking of the bottle holding the nuclear genie. Einstein and friends started to ponder their wisdom having seen 'progress' applied to Hiroshima and Nagasaki, and the many unsuccessful efforts to ban nuclear weapons bear testimony to the sea-change in the understanding of what progress is – the discontinuation of nuclear power in Germany is a concrete example of how an extravagantly prosperous country decides to abandon one of the fruits of human ingenuity.

P. Hulsroj, *What If We Don't Die?*, Springer Praxis Books,
DOI 10.1007/978-3-319-19093-8_3

Stopping progress??

But truth be told, there are not many examples of this kind and when progress without too obvious side-effects is made it is hard to believe that it will not be made available to populations whose most existential fear is to lose existence.

Beyond the secular domains of nuclear weapons and energy, religion provides some pickings in terms of halting alleged progress. The second strongest human impulse, lust, has been, and continues to be heavily circumscribed. Innovations in contraception are still frowned upon by the Catholic Church, and Ireland duly prohibited these immoral remedies until 1980. Islam is no fan of money-lending with interest no matter how useful for developing the economy. Christian sects have prohibitions on blood transfusions and all manner of standard medical procedures. Many kinds of recreational or transformational drugs are or have been outlawed, including alcohol. All of which goes to show that conviction can stop seeming progress – and can lead to societal disapproval in its various forms. The morality of immortality might not just be a question of individual morality. But it will also be a question of individual morality, and it will be a question of whether all this individual morality will aggregate into societal morality – and corresponding societal normative regulation. When looking to the Bible for inspiration one can look to Genesis – but also to the apocalypse that might be the ultimate price to pay for untamed hunger for the apple of knowledge. We might not want to rush there!

There are good reasons to be cautious when addressing how moral judgment informs societal normativity. Surely some form of morality is the foundation of all law; even hard-line positivists must admit this. Utilitarian considerations often lead to law or to abstention from law, but also utilitarianism is, of course, a morality. Yet to be cautious regarding the nexus of morality and law is justified, as so often in the past morality has lead us down the garden path. The ultra-libertarianism expressed so elegantly, and wrongly, by Robert Nozick, and so crudely by Ron Paul, is an extreme reaction to the tyranny of morality. The ultra-libertarians would argue against the morality of addressing the morality of immortality. As your neighbour's business is none of your business it follows that if your neighbour can achieve earthly immortality then that also is not your business – your neighbour can choose for himself. One could perhaps wish that it would be that simple, but it is not. Even if one believed in the most stripped down societal structure the question of who qualifies as the neighbour who should be left alone cannot be avoided. Are future generations neighbours whose path to existence can be blocked because an earlier generation decided to occupy the available space forever? Is it alright that they never get the chance to be left alone, because an earlier generation was left alone to decide; and decided in favour of itself. Of course, any procreation decision has this dimension to some extent, but when seats are taken forever and procreation consequently stops entirely even ultra-libertarian logic is moved beyond its original theoretical foundation.

What libertarians and other laissez-faire devotees logically miss is that when a freedom is exercised it always affects the freedom of others. If earthly immortality is chosen by individuals it impinges on the freedom of others to live between mortals only. When a child is not born the freedom the child would have enjoyed had it been born comes to naught; when future generations are not born their collective prospective freedom is extinguished without ever having come into being. There are those who claim that only beings that are born and can feel and plan should enjoy protection. However, that logic taken to its extreme means that both history and future lose significance, only the life of the moment has relevance. Few think that way, however. Most find historical truth to be important, and that honourable persons of the past should not be slandered; many worry about their sperm-count because a future descendant is important. In fact, assuming that future generations hold no rights runs counter to our evolutionary conditioned nature, where the A and O is passing on our genes. Evolution assumes our readiness to make the ultimate sacrifice in order to ensure the survival of our genes. If this is the case, how can it be argued that our genes in their future setting do not hold rights, perhaps even higher than our own? Polemically perhaps it can be asked how it can be the decision of individuals to stand in the way of evolution when evolution assumes unbroken continuity and every generation is a product thereof.

Chapter 4
What Is Eternity?

It has jokingly been said that sitting through an opera by Wagner is a taste of eternity. It starts and never seems to end. But, all disrespect for Wagner aside, how do we get our minds around the concept of something having a definable start and, per definition, no end?

In the concept of eternity there would naturally seem to be an assumption not only of no end, but also of no beginning. In cosmology we struggle with the idea of something arising out of nothing, yet in our personal lives we embrace easily the idea that there was a time of our non-being, while we struggle with the idea of our **future** non-being. Lucretius was the first to point out this asymmetry, unconvincingly explaining that the asymmetry should make us less concerned about our future non-being.[1] In cosmology we assume finiteness at both ends of existence, but in our lives we hope for the asymmetry of finiteness in creation and infiniteness in existence. That is the Christian Weltbild. For Hinduism and Buddhism the belief is generally more conceptually balanced,[2] with ambiguity galore, but with a tendency towards no start and no end. But in any event the idea of rebirth makes the question of finite versus infinite more abstract than when *one* death is assumed.

In Western thought Wordsworth has given voice to pre-existence, albeit in a rather undefined fashion:

> Our birth is but a sleep and a forgetting:
> The Soul that rises with us, our life's Star,
> Hath had elsewhere its setting,
> And cometh from afar:
> Not in entire forgetfulness,
> And not in utter nakedness,
> But trailing clouds of glory do we come

---

[1] De Rerum Natura.

[2] As, indeed, it is in Mormonism, where all souls are deemed to be co-equal with God and hence with no beginning and no end.

P. Hulsroj, *What If We Don't Die?*, Springer Praxis Books,
DOI 10.1007/978-3-319-19093-8_4

From God, who is our home:
Heaven lies about us in our infancy![3]

Or more prosaically stated by the band Shu-Bi-Dua in the iconic Danish song The Red Thread[4]:

What on earth are you
Before you become something
A shooting-star
A jigsaw puzzle?
A cock-up?
A carrot?
Hell's bells you must have been somebody!

...

So what's in store
When you've got to go
When the shirt of life
Is just simply too short?

Where does it lead,
The red thread?
Must lead somewhere, to something...

When matter is likely to be mirrored by anti-matter, and gravity is the clearing-house between the effects of different masses, one can start to speculate whether an expansion of the relativity theory does not entail that something must always be matched by a countervailing something else – that the human or cosmic sum of happiness may always be off-set by a human or cosmic sum of unhappiness.[5] On that logic, boundaries at one end of life must be off-set by boundaries at the other end, lack of boundaries at one end by lack of boundaries at the other.

[3] Intimations of Immortality. It should be noted that Wordsworth's idea of pre-existence is the opposite of what most people seek in immortality. For Wordsworth you arrive as a newborn as an unwritten sheet of paper, without experience, and exactly because of that lack of experience the child is 'the greatest philosopher'. The child is able to feel without the shackles of experience. However, what humans seek in immortality is to be able to carry forward the shackles defined by Wordsworth. The shackles of experience and the mature personality is what humans want to retain forever – not to be brought back to original matter or to a state of innocence wiping out individuality.

[4] Den roede traad, translation Harry Eyres.

[5] Clearly not on the individual level, however, since it seems obvious that some people lead happier lives than others. Yet, this does not exclude aggregation of happiness and unhappiness within a species, all living things or the cosmos – and a zero-sum balancing.

Thinking about boundaries or lack of boundaries is rooted in a perception of everlastingness; a concept difficult to reconcile with religious thinking about eternity, because it ultimately leads to the question of what the purpose is of eternal life. Can the purpose really be the eternal recurrence agonised over by Nietzsche, and if so is it truly a reward?[6] When earthly immortality is achieved that will be one of the ultimate dilemmas. Religion has another option for immortality, however, and that is atemporal existence – which is why Pascal's Wager remains relevant even when everlastingness of earthly life has been seized. Atemporality dispenses with both the questions of boundaries and lack of boundaries, because atemporality means that a being is outside time. Atemporality is the philosophical or religious one-up on Einstein and his space-time, although it has much older roots, Plato and his aionios, in fact.[7] Now, if atemporality can be conceived of at all, and that is debatable, then it might mean that everything exists simultaneously. As has been said, atemporality can mean that Nero is fiddling and Rome burning while someone is writing about Nero fiddling and Rome burning. Atemporality might not allow reversing of time because time is of no consequence, and hence it can, of course, be asked what the reward for a non-God is in having everything on the table simultaneously. Even more boredom? Probably not, since boredom is eminently time-bound. Pierre Gassendi was perhaps right to doubt that anything or anybody but God can be atemporal.[8] But, if that is so, how can we be God's children?

[6] Nietzsche was clearly in two minds about this. His amor fati suggests that the 'higher being' might want to change nothing in the lived life, and thus might welcome endless repetition. This is incongruent. If there is no continued consciousness both a repeated imperfect life and a repeated perfect life are as livable as it was first time around. And if there is some kind of continued consciousness then even a perfect life will lose its luster by eternal repetition.

[7] Plato, Timaeus: 'Wherefore he made an image of eternity which is time, having a uniform motion according to number, parted into months and days and years, and also having greater divisions of past, present, and future. These all apply to becoming in time, and have no meaning in relation to the eternal nature, whichever is and never was or will be; for the unchangeable is never older or younger, and when we say that he 'was' or 'will be,' we are mistaken, for these words are applicable only to becoming, and not to true being; and equally wrong are we in saying that what has become IS become and that what becomes IS becoming, and that the non-existent IS non-existent…These are the forms of time which imitate eternity and move in a circle measured by number'.

[8] Fifth Objections.

The perfect life is worth re-living endlessly?

Soeren Kierkegaard would appear to try to have it both ways when he suggests that God is living outside time and, yet, is present. However, as atemporality is a higher order concept compared to temporality there is, perhaps, no conflict. Atemporality might appear to allow taking on temporal presence, as done most explicitly with Jesus as the son of God, being part of time and even suffering in time. Kierkegaard's postulate becomes more difficult when transferred to ordinary humans: 'A human being is a synthesis of the infinite and the finite, of the temporal and the eternal, of freedom and necessity. In short a synthesis'.[9] 'The instant' according to Kierkegaard is the point when the eternal touches time. 'The instant is that ambiguity in which time and eternity touch each other, and with this the concept of *temporality* is posited, whereby time constantly intersects eternity and eternity constantly permeates time. Only now does the aforementioned division acquire its significance: the present time, the past time, the future time'.[10] So immediacy is eternity (more on this in Chap. 9), but this is not the same as humans having been made of both time and the timeless. Kierkegaard's view resonates with Christianity in the sense that humans are said to carry the seed of atemporality in them. It is, however, going far to suggest that this means that humans carry both qualities at the same time or that the two qualities have synthesized. One may question whether there is a point to such a logic, or lack of logic.

In any event, atemporality as the ultimate prize for beings that are steeped in temporality would not only appear paradoxical, but incongruent; a problem for Christianity with Jesus being both God and human and thus potentially both atemporal and temporal, the latter element brought out so poignantly in the Passion of Christ. Humans are physically and emotionally centred on temporality, that is the heritage of evolution and an integral part of being human. Removing that dimension would again seem to alter who we are beyond recognition, in fact, far more than going from water buffalo to dove. If continued identity is what is at stake then atemporality is the unlikely answer. A temporal being becoming an atemporal one seems contradictory, unless a metamorphosis of personality is the accompaniment, and then we have lost the plot.[11] It is easy to talk about the synthesis of the being, except if one has to explain what that synthesis is. Of course, it might be like boiling water and being left with salt and calcium, perhaps our sense of time is just the H_2O that disappears into eternity, yet few would argue that it is the salt and calcium that define the identity of the oceans in which we swim so comfortably.

But the real upshot is this: if atemporality exists will it not be intolerant of the everlasting, of sempiternity? An atemporal God might destroy sempiternity, but even a Godless perspective might lead to the result that atemporality and the everlasting are mutually exclusive. Not because atemporality is superior, but because

[9] The Sickness unto Death, 43 (Penguin Classics).

[10] The Concept of Anxiety, 108–9 (Liveright Publishing Corporation). This has similarities to McTaggart, explored below at footnote 3 at Chap. 9.

[11] Somewhat in that direction Bernard Williams, The Makropulos Case: Reflections on the Tedium of Immortality, in David Benetar, Life, Death & Meaning.

everlastingness is still time dependent, and if something is time-dependent then there is hardly any assurance of forever. Only atemporality is demonstrably eternal, only atemporal life is eternal life, because it is unconditioned, does not depend on external factors like time.

These distinctions are relevant for the purposes of this book in so far as we assume that earthly immortality is temporal, and therefore inherently indeterminable as truly everlasting. In the right circumstances life might be everlasting, but in the wrong circumstances it might still be extinguished – or death might be willed. Immortality is a possibility – never a certainty in the earthly and time bound variant! Only the atemporal can jump off the cliff and not die a definitive death!

Chapter 5
The Timeless and the Ageless

When something is being categorised as timeless in this book the meaning is that something is atemporal. Conceptually, timeless can also refer to something which has no reference to a setting in time, however.

When the meaning is that something is independent of setting in time, like when Rod Stewart sings about 'timeless, ageless beauty'[1] we start to struggle. Is Rod Stewart suggesting that his lover would have been beautiful also to the Neanderthal? Not so obvious, and not obvious at all that any feature of beauty can remain untouched by time. The Cro Magnon might have demonstrated a nature aesthetic not dissimilar to our own in their cave paintings, but what about our aesthetic before evolution made us human? Our nature appreciation might well have been rather different when we made the first move from water-being to land-being.

Be this as it may, 'timeless' and 'being' make for uneasy partners.

And it does not become better with 'ageless'. In a love song like that of Rod Stewart one would hope that the idea is not that the lover comprises all ages at the same time. But if not that, what is then the meaning? Ageless as in anodyne? Certainly not. Perhaps a transcending beauty that cannot be corrupted by time? Perhaps, yet Stewart attributes in the same line 'lace and fineness' to the love interest. So a juxtaposition of the most constant with the most ephemeral – 'timeless, ageless' on the one side, 'lace and fineness' on the other. Surely a rocker like Rod Stewart has not deliberately constructed a Faustian dilemma, but that is, in a sense, what it is. Faust sought to embrace the ephemeral for a 'timeless' moment, although he knew that the price he had to pay for this was death and everlasting servitude to the devil. Are we hoping for one without the other? If we do, then Rod Stewart's song is just one small piece of evidence of how hopelessly mixed up we are when thinking of immortality.

We might associate immortality with being timeless and ageless, but the language itself shows that this can only make sense in atemporality. You cannot be timeless and ageless in time. Still, we might understand ourselves as a diamond

[1] 'You're in my heart'.

P. Hulsroj, *What If We Don't Die?*, Springer Praxis Books,
DOI 10.1007/978-3-319-19093-8_5

filter, unmoved and unchanged by the light of experience that passes through, but that is hardly distinguishable from death, and in any event only something imaginable in the beyond. If we achieve immortality on earth it would not, could not, should not make us timeless and ageless. Living is an evolving quality, as is beauty mostly. We might find comfort in the unchanging beauty of the hare with the amber eyes, because the hare becomes a symbol of constancy when in juxtaposition with the convulsions of the life around it. And hence we might imagine ourselves as the hare – the constant identity contrasting the ever-changing life around us. But, of course, that is tantamount to standing still – to be truly dead.

In Janacek's famous opera, The Makropulos Affair, Elina Makropulos has gained a timeless, ageless exterior beauty, frozen at 42, whereas her soul ages, growing ever colder, ever more unimpressionable. After 337 years the tension between the apparent timelessness and the actual age brings the edifice to collapse, with frighteningly rapid physical aging and ultimate death as the result. If humans in the future can seize earthly immortality they will also seize the Makropulos tension, and, as discussed later, might well embrace death as readily as she does when they have had their fill of life, however long that might be.

Faust, really?
(Credit: Allen Warren)

Chapter 6
Time and Existence

What we strive for in immortality is the continuation of existence. And existence is a temporal concept, which can be brought in line with everlastingness, but which might be ontologically incompatible with atemporality.

In previous chapters the tension between human nature, bound to the temporal, and a possible embrace of the timeless was explored. However, language, as an expression of how we think, might also show how we have built a barrier between ourselves and atemporality.

The word existence, emanating from Latin and adopted in almost unchanged form by most Western languages, implies 'emergence' or to 'stand alone', which again implies that we have emerged from somewhere or something else, or that we stand alone, where before we did not! Kierkegaard argues that through existence we have left the divine timelessness, God's 'is'. Existence then, by definition, involves time, or Plato's 'imitation of time', 'a uniform motion according to number'. Existence becomes antithetical to timelessness, and the kind of existential immortality we hope for can only be realised as earthly or celestial sempiternity. Of course, we can also speculate on giving up existence, strive to reverse the 'emergence' involved in existence, and go back to the divine 'is'. After all, if we can emerge from timelessness, we should also be able to remerge and attain the earlier state of 'is'. Since this means losing individuality it would, however, seem to be contrary to the interests we have defined for ourselves with immortality – it gets close to the pantheistic conundrum which will be addressed in the next chapter.

It can be argued that a juxtaposition of existence and atemporality is just a language dispute, dissociated from any substance. This may be so, yet language is a mirror held up in front of our dreams and passions, and we all understand the word 'existence' to involve time, even if we disregard the origin of the word. To muddle our minds and, perhaps, to attempt to allow us to have our pudding and eat it too, we use time-associated language even when we describe what we might really believe is timeless. 'Does God exist?', 'God is dead!' are statements of that character.

P. Hulsroj, *What If We Don't Die?*, Springer Praxis Books,
DOI 10.1007/978-3-319-19093-8_6

Yet, many will believe that the question should be framed in timelessness terms.[1] The fact that a statement like 'Is God?' or 'God is not!' is linguistically extremely painful,[2] shows that our whole conceptual universe turns on time, as Kant correctly pointed out. In physics there is, nevertheless, no lack of theories on timelessness, Einstein's space time and time as a fourth dimension being the most prominent, closely raced by string theory. Yet, no physicist has suggested that the Big Bang was the consequence of a move from the timeless to the temporal,[3] although this might have as much credence as many other theories. Perhaps the Universe, just like humans, emerged into time from a timeless abode – time and the timeless thus living side-by-side. Physicists might try to escape this discussion by talking about the singularity, the domain where physical laws break down. But, of course, it could be the other way round: our physical laws might be the result of the breakdown of a more perfect order, that of timelessness. The Big Bang that gave birth to the Universe might be a chip off the block of atemporality and the Big Crunch might be the return to the mother lode, with many chips of the atemporality block giving birth to a plethora of universes and many Big Crunches bringing many returns of time-bound universes to atemporality.

In a similar fashion our birth might be our private Big Bang taking us out of atemporality, and our death our private Big Crunch – perhaps we have a private cosmology which mirrors the universal or multiversal cosmology. All things time-bound leaving the bosom of atemporality only to return later!

If time has arisen from atemporality it can then be argued that also we are derived from atemporality. With that assumption two options exist, one, that each human being is derived from atemporality, but has no atemporal identity, and, the other, that we as individuals descend from atemporality to time, only to return as individuals to atemporality when our time is up. The latter possibility goes far beyond pantheism or even Wordsworth because our individuality is retained, when we as a chess piece are taken out of the box to play in timely existence, only to go back to the box upon the end of play, yet still being the knight or pawn we were when we came out of the box and which we were while we played in time. Perhaps we have played many times, Hinduism/Buddhism – like, and carry the experiences of previous games with us, out of even our own sight while we are on the board.

The difficulty with this line of thought is that we do not understand, probably cannot understand, what it means to be the knight or the pawn in the box. As we play in time, we play under the premises of time, and identity as something timeless cannot be grasped, possibly should not be grasped. It is Kierkegaard's leap of faith

[1] In this fashion Soeren Kierkegaard, 'God does not exist, he *is* eternal'.

[2] Heidegger in Being and Tine does what he can to demonstrate this painfulness: 'Trees are, but they do not exist' and equally 'God is, but he does not exist'. For Heidegger the point is not just the time-bound element, however. Existence means potentiality-for-being, and although there is a time element to this, this is not sufficient for existence, as Heidegger's refusal of granting existence to the tree shows.

[3] Although Hawking comes close when he talks about the atemporal as the possible status before the Big Bang. However, his atemporality seems to be a void, not an 'all-at-the-same-time'.

in some sense. This book assumes that our interest in immortality is to preserve our personality and our experiences, yet atemporal identity might, or might not, allow this. On the one hand, this may be one of our eternal Schroedinger's cat mysteries, although there is beauty in the thought that individuality in timelessness might be similar to holding all experiences, sorrows and joys, in a diamond-like indestructible and unchanging immediacy, giving ultimate truth to 'eternally owned is but what's lost'[4] (except, of course, that it is not lost in that case). On the other hand, the reality might be that we are not able to formulate the question on eternal identity that appears so essential to us, because we do not know what answer we hope for when the prism is atemporality.

The reason that atemporality and its consequences are not discussed more is probably, again, that the human mind and the language it produces struggles to understand that something can be without existing. In a strict interpretation of the word 'eternal' it means timeless.[5] But when we dream about eternal life, of what do we dream? We dream of 'exist', but using 'eternal' makes the statement so comfortably ambiguous. It makes us avoid reflecting on what it actually means. The distinction between 'exist' and 'be' is one we almost cannot make, conceivably because it could make clear to us that it is perfectly possible to be and yet not to exist!

In a sense there could be consolation in this. Perhaps our private Big Crunch is, indeed, a return to the unity of atemporality, a return to a timeless moment holding all experiences of all universal or multiversal existences, and perhaps even of all possible experiences of all possible multiverses and all their beings and matter – the latter being a supercharged version of the concept of parallel universes, but bringing yet another fundamental question as to whether in atemporality there is an ontological difference between reality and possibility, between you having existed and you having existed as a possibility.

In this plethora of basic questions the most confounding one might ultimately be whether in atemporality of this nature you retain personality or whether a return to the unity of atemporality means that individual identity is lost. In keeping with the all-inclusiveness of atemporality the quantum mechanics inspired answer might be that both are true: in atemporality you might at the same time retain and lose personality! Even more radically, quantum-mechanics logic might lead to the conclusion that we, with death, will both be and not be; will both be and exist and not be and not exist! And completely mind-bogglingly, this might be true even before death. Yet, on earth we just cannot 'break on through to the other side'.[6]

[4] Henrik Ibsen, Brand, Act IV.

[5] In the rest of this book eternal is used in the common meaning, thus covering both the timeless and the everlasting, as it is, indeed, convenient to have a term that covers both.

[6] The Doors.

Chapter 7
Pantheism and Spirit as Substance

There is a certain beauty in the idea that everything lives, that everything has spirit, that everything is God. This beauty has special appeal to those who have difficulties with a personalised concept of God. The belief of some Native Americans that the spirits of ancestors live as part of Nature is a pleasing blend of abstraction and emotional continuity. Yet, the belief is baffling to the extent that the belief is that identifiable spirits occupy Nature, although it becomes very aligned with physicalist thinking if the idea is that the spirit of ancestors becomes part of the spiritual substrate which 'animates' all and everything. If the spirits of ancestors lose individuality and are merely recycled into the Great Spirit then the belief has resonance, but has lost the reassuring quality we generally seek in immortality because individuality has been lost. How does it help our immortality to know that the substance of our spirit lives on, if consciousness is lost?

This is not to deny the beauty of pantheism, with its many different flavours, only to say that from the perspective of the dead there is no appreciable advantage in being part of the Great Spirit, if what you were hoping for was continued individuality. Or not more advantage than the immortality derived from fame, a la Sartre, as will be discussed below. The attraction of pantheism in the sense discussed here is extra-sensory and non-individualistic. Or, the attraction is for the survivors,[1] who might find great comfort in knowing that the spirit of ancestors permeate their environment, assuming that somehow the qualities of the spirits of the deceased will play out in Nature, even when individuality has been lost.

If the human interest in immortality is an interest in continued individual consciousness there is little solace to be found in pantheism understood as collective existence in the Great Spirit. The delineation of the You relative to the Other might disappear – a beautiful thought – but this only leads to dilemmas similar to those of collective human consciousness (also discussed later). To the extent that a 'You' feeling element is imaginable in collective human consciousness then that 'You' element will be overwhelmed by the equally ranked billions of 'Other' feelings in

[1] Like for resurrecting the DNA of parents, see Chapter 40.

P. Hulsroj, *What If We Don't Die?*, Springer Praxis Books,
DOI 10.1007/978-3-319-19093-8_7

the collective abode of the 'Great Spirit'. Overwhelmed to such an extent that the You might hardly be recognised, which in a sense is exactly the point of the pantheistic viewpoint. We come full circle![2]

The idea of retention of individuality is powerful and even pantheism does not always completely erase it. The consequences of understanding spirit as a substance has been only dimly acknowledged. The perhaps uncomfortable truth of spirit as a substance is that spirit may never disappear, but it may permutate into many different forms without retaining the spirit of an individual as a continuously self-standing entity. Often you hear statements like: 'if I turn into a stone when I die', the assumption thus being that your spirit will stay together. You never hear people say: 'what if I turn into a quarter of a lettuce head, a tenth of a rain-worm, a fifth of a granite stone, ….', the implication being not only that your individuality disappears and your spirit is divided, but that your spirit will join the spirit substance of others to create new self-standing entities of spirit. That is the radical take on rebirth; that, unlike in Hinduism and Buddhism, you do not stay intact. The release to be found is then not from desire, but from individuality![3]

[2] This discussion of pantheism does not speak to the issue of monism versus dualism; it addresses merely the narrow perspective of the interest in immortality being motivated by the desire for continued consciousness.

[3] As may be the case also for Hindus and Buddhists, not in rebirth, but when Nirvana is reached.

Chapter 8
Communitarian Happiness

This book assumes that our desire for immortality is a desire for immortality of our individuality. We hope for eternal happiness, and we assume that we cannot be happy if we do not know that we are happy. Individuality is our tool to internalise feeling, and without this tool, what is the point? Our constituent parts might be happy in other configurations, since we believe with the natural sciences that nothing perishes, but is the role of our 'self' really just to be a gracious host to our constituent parts, and then with age and death to fade away? Does the destruction of the self not contradict our belief that nothing perishes, even if Hume tells us that our self perishes every moment, which, of course, we also do not believe. In the final analysis the fundamental issue is perhaps whether our self is just itself an amalgamation of constituent parts or whether the self is an indivisible building block which either perishes in its entirety or is indestructible in its entirety. While this may sound like a religious issue, it is, in fact, more than that.

Locke explained to us the difficulty in identity terms of making sense of the leaf, the branch and the whole tree, and in a similar fashion one can ask why we assume that happiness resides in the self, the whole self. Over the centuries there have been theories on the soul residing in the heart, in the digestive system, in the brain, or, as advocated by Descartes who was never afraid to go out on a limb, more specifically in the pineal gland. So clearly it is easy to be mistaken in questions of this nature, and would it not be tragic if we strive for immortality on the assumption that the self is the home of happiness, only to find that this is not at all the case?

Socrates talked about the blessing of death as similar to the dreamless sleep, and, indeed, when you wake up some mornings after wonderful sleep you can notice how many of your constituent parts signal satisfaction; the muscles being relaxed, the heart being quiet, and entering the bathroom you might see skin and hair being unusually attractive. So your dreamless sleep has made many of your stakeholders very happy. Of course, they were equally happy, or even more so, before you woke up, so your self is only important for them as the organising principle allowing them to be happy – another organising principle might make them equally happy. Conversely, your great toe, which is not an essential part of your self, might make

P. Hulsroj, *What If We Don't Die?*, Springer Praxis Books,
DOI 10.1007/978-3-319-19093-8_8

you intensely unhappy when you slip and cut it on the treacherous stone on the beach. The message of your great toe might be understood to be, that you, as its organising principle, should hurry up to make it happy because it will otherwise eliminate you as its organiser, and will mutate into a more auspicious form. Dualists might find this to be confirmation that there is a difference between who we are and what we are made of and even if this might be true it is possibly not the whole story. Just as Darwinists are telling us that our genes seek their happiness independently of us as their vessel, perhaps all our constituent parts are more interested in their individual happiness than in collective happiness. Nature tends not to be communist, and clearly our constituent parts are not either. The blood you spilled on that damned beach stone will not clamour to return to you as its vessel, it will transform and be happy in some other form. Just as Hume assumes constantly new personal identity, the reality is that our body is also constantly taking on new forms often outside us. So few are the molecules that follow us loyally from cradle to grave!

An ultra-pantheistic perspective might then bring the conclusion that our endeavour should be to provide as well as possible for all our constituent parts, just as we have been evolutionarily conditioned to do for our genes. Do not flush your clipped finger nails in the toilet if this is not their best abode! Even more radically, this perspective could also lead to humankind becoming far more fiduciary in the way it treats the environment in general. If we must do the best we can for our discarded physicality must we not do the best we can for all living things? For all things?

But what about our good old self, which might not be physical? Can it dissolve and join a larger identity pool on Earth or in the heavens? Individuality is premised on the finite. Perhaps ultimate happiness is boundless and hence incompatible with individuality?[1] Perhaps individuality is in the final analysis a mirage and perhaps our personality substrate can gain happiness without us having to be happiness score keepers. If so, it is back to Hinduism and Buddhism in a sense: the ultimate happiness might be the release from individuality! And the obvious conclusion from this would be that when ultimate happiness is the release from individuality we should not strive for earthly immortality. The book could end here. Yet the thought

[1] A bit in this direction Schopenhauer: Death is a sleep where individuality is forgotten. And also the inscription on the headstone of Erwin Schrödinger's grave is relevant:

Denn das, was ist, ist nicht weil wir es fühlen,
Und ist nicht nicht, weil wir es nicht mehr fühlen
Weil es besteht, sind wir und sind so dauernd.
So ist dann alles Sein, ein einzig Sein.
Und dass es weiter ist, wenn einer stirbt,
Sagt Dir, dass er nicht aufgehört zu sein.

Then that which is, is not because we feel it
And isn't nothing, because we no longer feel it
Since it is, we are and are so permanent.
So all Being is a single Being
And that it continues to be, when someone dies
Tells you that he did not cease to be.
(translation Harry Eyres)

of immortality is almost immortal in itself, many humans will continue seeing individual indestructability as the ultimate human accomplishment, and hence investigation of consequences, notwithstanding the possible irrelevance of the self, will be continued below.

Before doing so a look at a Christian perspective on identity might be warranted, namely the one where humans at the outset live in God only to become a fragment of the divine through individual life and, with death, return to live in God. This perspective has an uncanny resemblance to the one discussed earlier of humans being spun off from atemporality through existence and returning to atemporality with death. From the continuation of identity point-of-view the question then becomes one of whether the return to God extinguishes personal experience and personality or not. Whether the individual becomes a part of a greater whole or will be overridden by a greater whole.

However, in the final analysis such a perspective also means that humans are constituent elements of God throughout: while we live in God, while we gain terrestrial manifestation as humans, and when we return to live in God. The interest of the Über-ego of God in retaining us as individual constituents with preservation of identifiable personalities and experience, and our 'own' interest in retaining memory of our brief terrestrial moment, might be questioned, yet with God all-knowing, and us being part of God, perhaps the logic is exactly that nothing is lost, including our personalities. How relevant that will seem to us after the return to God is another matter, considering that we as part of God will share in all God's ability and knowledge.

There is also a non-theistic take on this. Many humans have a feeling that dying is a home-coming, and that life was an odyssey set out upon from a shore of Wordsworthian wholeness. That death will make us whole again.

This wholeness which we left and to which we return might be a super-consciousness. Dying from that super-consciousness, through becoming human upon birth, might have been as frightening to the splinter as we now perceive our deaths to be. Leaving the original shore we might have feared that we would lose our super-consciousness personality,[2] just as we now fear that we will lose our earthly personality when we go back to becoming part of super-consciousness. Becoming whole again through death might be the relief that the way back was ultimately found. Ironically, our super-consciousness might not put great stock in our terrestrial interlude. That part of super-consciousness that is each of us individually might agree!

[2] Barring all-knowingness, of course.

Chapter 9
Ex hoc momento pendet aeternitas

There is a more poetic understanding of eternity than merely whether it is atemporal or sempiternal, and that is one where each moment in an almost Kierkegaardian fashion contains both temporality and the eternal.[1] Most people have experienced moments where everything seemed to come together, and you almost wanted to shout with Faust: stay, thou art so beautiful! This feeling is one beyond the Hindi/Buddhist elimination of desire; it is a feeling of desire satisfied, 'the little death' of orgasm.

Rita Hayworth famously said that her men went to bed with Gilda, but woke up with her. The tragedy of that statement for her suitors is that they chased an essentially static dream of surface perfection, instead of communion with a lover's beating heart. Yet, only the beating heart will bring the feeling of eternity in a temporal sense: the lover's beating heart is what makes you transcend the self by joining together the temporal and the eternal. By opening up to the Other, the protection our personality provides against the enormity of universal forces is briefly pierced. The Gilda perfection is akin to the purely atemporal, it is lifeless, does not connect with time. Paradoxically only life – at its most intense – allows time-bound human beings a taste of timelessness.

[1] See text at footnote 9 of Chap. 4.

P. Hulsroj, *What If We Don't Die?*, Springer Praxis Books,
DOI 10.1007/978-3-319-19093-8_9

And the morning after?

One of the downsides of the objectification of sex is that it ultimately allows us to not open up to the Other, and thus we miss out on the most precious gift bestowed upon us: the ability to transcend. 'There is a land of the living and a land of the dead and the bridge is love, the only survival, the only meaning'.[2]

Of course, it is not only in love or in sex that things can come together in such a way that eternity shines though temporal experience, that the future is not in tension with the past and the present.[3] Sex and love are symbols of even more existential issues, no matter what Freud and Darwin have been telling us. We may have been conditioned in a certain fashion and we might not be able to resist that conditioning, but beneath all, of course, is the question of how I, as an atom or a neutron in a gigantic construction, can make sense of my individual existence. Some might seek anthropocentric answers, others might find none, but the human struggle at its most fundamental is about the conciliation of my individual insignificance and the universal enormity. The time when things come together in brief moments of ultimate balance is the time when our insignificance and the whole edifice come together, where the atom feels at home in its much wider context. Some people will argue that this is the most an individual will ever be able to live eternity!

Faust was ready to sell his soul for that one moment of perfect balance; thought that this one taste of perfection and eternity was worth everlasting servitude to the devil. In a very roundabout fashion you might argue that the perfect moment was Faust's experience of atemporality in time and that, for him, sempiternal pain was outweighed by this brief moment of ultimate balance.

There is a residual question of how perfectly balanced these moments of perfect balance really are. The stillness, which is at the core of the balance, is our taste of eternity, and yet are we perfectly still when all things come together? For Faust this moment was the moment of perdition and hence, perhaps, for him the balance was perfect. But when we feel that all things have come together, on inspection, perhaps, there is still an element of movement: the future or the past still pulling gently, soft tones accompanying the otherwise perfect stillness. So perhaps the point is that in these situations we get as close as is humanly possible to eternity. Yet, it might be suggested that if the gentle pull disappeared and we experienced truly perfect balance then we would die, because human life can be sustained only when its dynamic element is present, even if it is in the smallest of doses.

The fundamental epistemological question raised by ex hoc momento pendet aeternitas is whether the ability to transcend provides an element of proof of eternity and humankind's belonging to it. This would be similar to how the mere fashioning

[2] Thornton Wilder, The Bridge of San Luis Rey.

[3] This way of perceiving eternity is similar to the B series of McTaggart, where time is not seen from the inside and hence it is not possible to talk about past, present and future. In B series time is viewed from the outside and 'earlier' and 'later' are the only navigational tools, J.M.E. McTaggart, The Unreality of Time.

of concepts might provide proof of the possibility of a corresponding reality, or the more radical modal realism of David Lewis, who insisted that all possible worlds are as real as the actual world.[4] In an almost Cartesian perspective surely a strong feeling of the eternal is a piece of, inconclusive, evidence, but more importantly such feelings are highly persuasive in a dialectic which, despite all our science and alleged objectivity, remains a matter of appreciation and hence highly subjective. If the starting point on immortality is a 50/50 probability of truth, experience of eternity, no matter how subjective, pulls the odds in the direction of its reality. 'The insatiable thirst for everything which lies beyond, and which life reveals, is the most living proof of our immortality' is how Baudelaire put it.[5]

A thing can only be understood in juxtaposition to its opposite. Matter – anti-matter is not the proof, as matter could also be contrasted with no matter, one kind of matter against other kinds of matter. But if you are convinced about mortality you also find proof of the possibility of immortality[6] – in a logic a bit inspired by the young Wittgenstein[7] and adopted strongly by Hermann Hesse in Siddhartha:

> … in every truth the opposite is equally true. For example, a truth can only be expressed and enveloped in words if it is one-sided. Everything that is thought and expressed in words is one-sided, only half the truth; it all lacks totality, completeness, unity.

Hesse and Siddhartha then quickly move towards a rather Hegelian conclusion, however:

> But the world itself, being in and around us, is never one-sided. Never is a man or a deed wholly Samsara or wholly Nirvana; never is a man wholly a saint or a sinner. This only seems so because we suffer the illusion that time is something real. Time is not real, Govinda. I have realized this repeatedly. And if time is not real, then the dividing line that seems to lie between this world and eternity <Welt und Ewigheit>, between suffering and bliss, between good and evil, is also an illusion.

That humans possess both worldliness and eternity reconciles well with Christianity, with Jesus and humans being born children of both God and parents, and maps on perfectly to the idea of death as the portal to immortality. But even Hesse's extreme holism cannot explain a necessity of mortality having to be accompanied by earthly immortality. Hesse's assumption of truth in pairs might be pleasing in a juxta-positioning of Welt and Ewigkeit, not of Welt and weltlicher

[4] On the plurality of worlds.

[5] Even Lincoln has weighed in 'Surely God would not have created such a being as man, with an ability to grasp the infinite, to exist only for a day! No, no, man was made for immortality'.

[6] In this context the zeros and ones of computers are also worth noting, as they can reduce any proposition to pairs of alternatives. This reflects also the human thought processes where all comparison in the final analysis is comparisons of pairs. Even a comparison of three items will be reduced to pair comparisons. Hegel's dialectic puts a spin on this with the assumption that thesis and anti-thesis result in synthesis. Pair comparison might not prove the truth of any of the involved propositions but will prove the possibility of truth of both.

[7] 'The limits of my language are the limits of my world'.

(earthly) Ewigkeit! It is true that earthly immortality will always leave open the possibility of the mortality of elective death; the two must always co-exist. But as we have sadly learned over the millennia, mortality can exist perfectly well without earthly immortality as its mirror. And impossibilities might exist: on the earth you currently inhabit it is not easy to imagine that you can be dead and undead at the same time, no matter what Schrödinger and his cat might have suggested.[8]

[8] But see right below on possible modifications to the impossibility paradigm.

Chapter 10
Immortality Through Parallel Universes

One of the most far-out thoughts in quantum physics is the idea that every option leads to its own reality. If you can choose to go left or right and you choose right then a parallel universe opens up in which a parallel you chooses left. An immense number of yous then exist: a parallel you for every choice you have ever made, or which has been made by others but involving you. An almost infinite number of parallel yous consequently exists, each with their own life trajectory. And not only of yous, but of your alternative yous. All the sperm cells that ultimately did not make it to your mother's egg, will have made it in parallel universes. If you take this logic all the way back through multiversal history, ultimately all potentialities that ever arose will have sparked a chain reaction of parallel universes. Far-out, indeed!

It has been suggested that all these parallel universes represent a kind of immortality (immortality squared in the extreme, in fact), since another you will always live on, even if the *you* you dies. Yet this is debatable even in this extreme logic, because aging is not a choice, and hence all the yous will ultimately die. Bad choices will allow the good choice yous to live on, even if the bad choice you dies, but this does not amount to immortality. The inevitability of the laws of nature will still kill all the yous, unless you pursue the further radical thought that immortality could have been invented (as is the assumption of this book) and that therefore a string of yous has arisen from the possibility of immortality. And those yous would then all be immortal. But, strictly speaking, this does not mean that the parallel universes have given some of the yous immortality. It is the possibility of immortality that has given a cascade of yous immortality but, sadly, not the *you* you!

And with the lack of immortality for the *you* you lies the rub! Many yous may exist, as also proposed by Derek Parfit, who does not even need to resort to parallel universes to theoretically divide a self into several selves; tele-transportation being his main tool of hypothesizing. The many yous spawned by parallel universes or tele-transportation share an identical history up to the point of their separate individual creation. Thereafter the other yous become less and less you because they experience their own future which is different from that of the *you* you, even if the past is shared.

P. Hulsroj, *What If We Don't Die?*, Springer Praxis Books,
DOI 10.1007/978-3-319-19093-8_10

Infinite Jimmy Hendrix

The fundamental problem with immortality in this constellation is, however, that each you, identical or not, wants to be immortal, since each has consciousness and each wants that consciousness to continue. It does not help your immortality that another you becomes immortal. You want it for yourself. Immortality for your other self is similar to immortality through progeny – but it is not *your* immortality. If your wish is to gift some worlds and their humanities with your ability and your particular DNA combination then parallel universes (or cloning) might be the answer, but if your wish is to gift yourself immortality parallel universes will not do the job!

In his brilliant book 'Our Mathematical Universe', Max Tegmark explained a thought experiment on quantum suicide that illuminates the dilemma of the many yous. The basic premise is that quantum mechanics lead to all choice being matched by its alternative in a parallel universe when quantum superposition (something being in two states at the same time) is achieved relative to the choice. Hence a version of you will always live on if you devise a suicide trigger mechanism (a quantum machine gun) that is based on the achievement of superposition a fraction of a second before a gun is firing or not depending on whether it is measured to be in one or the other state. In this scenario, a version of you will always live on because of the parallel universe created by the state of superposition, since in superposition the gun would both fire and not fire. And this is true even if the cumulative probability of survival is zero if the suicide mechanism is endlessly triggered and you were subject to the normal laws of probability. Tegmark's quantum suicide experiment is a variation of the famous Schrödinger's cat experiment, where in quantum theory one version of the cat will always survive, unless you believe in quantum collapse.

Tegmark's logic is impeccable except that it overlooks the interest in survival of each version of you. It overlooks the continuity of consciousness element. In his experiment a large number of yous will die, including, with a very high degree of certainty, the *you* you. The you who reads these lines will die, even if identical copies of you, each with their own distinct consciousness will live on. Tegmark's assumption of your quantum immortality is flawed not only because of the inevitability of aging, but because the you with your consciousness will die. You might find little consolation in the fact that an identical you, with its own consciousness will live on.

That there is a *you* you, a unique self for you that carries your unique consciousness, does not imply that your reality is the original reality and that all the parallel yous are less real or less original. The caution of Ernst Bloch against this sort of distinction continues to apply. All that the *you* you implies is your understandable interest in the continuation of your specific consciousness. And Max Tegmark's experiment fails miserably in this respect.

What is more, all the yous except one will die in the suicide experiment, if the you that survives the first firing is immediately exposed to the next firing and so on. At the end of each firing only one you will be alive, and hence a string of firings will leave a string of dead yous and always only one surviving you at the end. This might not worry the *you* you, because the *you* you in all likelihood would have died early in the string of firings. But in any event you would only care about all the dead yous if you believed that all yous are connected through a kind of Über-Ich, in which case you would share in all delights and tragedies of an infinite number of yous, something not even quantum theory implies, but which can be argued to resonate somewhat with Hindi/Buddhist thought on re-birth and the sharing of a mother lode.

The tragedy of the parallel universe concept, or perhaps the relief, is, that although an infinite number of yous and alternative yous exists, you as you are on your own, sharing nothing with all your alter egos, and having no comfort from being one of an immense multitude. All you care for is the you that you live and that you are continuously conscious of.

Yet, parallel universes pose an empathy problem. All your other yous are similar to brothers, so when they suffer it is similar to your brothers suffering; distant, invisible brothers, it is true, but still brothers, many of whom even share your DNA completely. Identical twins. So should you not care? Every action or inaction by you will give rise to the alternatives for your brothers, and sometimes the alternatives will be terrible. And you are caught, because whatever you do or do not do will propagate in the parallel universes. There is no way to avoid the pain in the parallel universe – or, more positively, to stand in the way of all the alternative pleasure. What is more, your off-spring will not be only the children known to the *you* you, your children will also propagate throughout an infinite number of universes, living though all the pleasure and pain conceivable, and you do not even have the choice to not have children, because in parallel universes parallel yous will have the children. One parallel you will have one child, another two children, yet another twenty. For you the only question will be whether the *you* you will procreate and to what extent.

One important conclusion would then seem to be that the empathy issue cannot be resolved, particularly as long as we do not know for sure whether parallel universes exist, but that parallel universes could turn our value systems almost completely upside down. Another conclusion, of higher certainty, is that you have no choice but to make the best choices for the *you* you and the progeny of the *you* you. It may be argued that it might not matter because other yous will in any event live your discarded choices, but if you do not try your very best for the *you* you and your dependents you will introduce a bias into the equation a la fatalism (and, yes, therefore this will also be lived). But this bias would count against the *you* you, because you are the possessor of a specific continuous consciousness and hence should have an interest in obtaining the best results for this specific you. There is a difference between the *you* you and the Other. When you do not partake directly in the lives of your other yous, your interest will be optimisation for the *you* you. Possible empathy cannot logically lead to indifference towards the circumstances of your specific you! Except in a parallel universe!

Evolution logic is, of course, also challenged by the concept of parallel universes. When bad choices are forgiven in a parallel universe and the weak survive in their universes alongside the strong in theirs then it is perhaps only our universe, and its parallels, that adheres to the laws of evolution. Or, perhaps there will be universes with stronger implementation and universes with weaker implementation, all shades of the evolution principle being played out in some universe, and a vast multitude of universes where one, several or many concrete exceptions to evolution will have come to pass. If evolution is only one of the effective laws of physics rather than one of the fundamental laws, then you would expect that parallel universes embrace also all the alternatives to evolution, and all alternatives to the concrete results of evolution. The survival of the strong and the death of the weak being matched somewhere by the death of the strong and the survival of the weak, and somewhere else by the death of the strong and the weak, but the survival of the middling. The logic of this strand of quantum mechanics is, after all, that all states exist, even the ones that give lie to the effective laws we hold dear and consider fundamental. The limits of the real might be the limits of the possible!

Chapter 11
Death of Compassion?

A lot will be said in the following about what society should and should not do in terms of the pursuit of earthly immortality. Yet, with the possibility of countless parallel universes the question arises whether we should be concerned about society's reactions to eternity, at all. In radical multiverse theory all possible scenarios will play out, wonderful or terrible, so why worry about society's direction in your universe except when you are personally and directly emotionally engaged; why worry unless the wellbeing of you and yours is involved? Immortality for you and yours might be interesting, but society's general attitude might be uninteresting. With society and all other beings outside your immediate circle you are in any event in an abstract relationship, in which you are not more invested than in the relationship with parallel universes and their actors. If you are Swedish the Paraguayans in the universe you share are, in a sense, as remote for you as the Swedes in a parallel universe. Hence, the human beings you do not meet in your universe are as much strangers as the parallel yous in parallel universes are strangers to you. The other yous will meet all possible ends, be they sweet or bitter, in parallel universes, and similarly every Swede will meet all possible ends in parallel universes, just as Paraguayans who meet a good end in your universe will meet a number of bad ends in other universes. If you ensure a good life for them in your universe you only shift the bad life for them to another universe. If immortality is achieved in your universe the result is that it will not be achieved in another.

The consequence of this perspective might then be that you should love your neighbour, but only if he or she is truly your neighbour, is truly close in an emotional sense. The fatalism rejected in the previous section for the conduct of your own life you might instead apply to those with whom you have an abstract relationship only. Genocide in a remote corner of your universe is of no concern, because if you successfully fight it in your universe it will take place in another universe. All that matters is what happens to you and to those in whom you are very directly emotionally invested. Proximity becomes all important for the exercise of your free will.

Now, this might be an objectionable theory merely because we do not know whether radical multiverse theory represents reality. You cannot let the world around

P. Hulsroj, *What If We Don't Die?*, Springer Praxis Books,
DOI 10.1007/978-3-319-19093-8_11

you suffer on the assumption that a given theory is correct. Even so, the question is whether such an approach to ethics would be logical and acceptable if we knew that radical multiverse theory was accurate.

Peter Singer has suggested that our moral obligation must be the same regardless of whether someone is close to us emotionally or not. This is a beautiful concept, which is, however, entirely Weltfremd, and which relative to radical multiverse theory would, indeed, lead to those suffering in a parallel universe commanding the same compassion as those we love in a more direct and concrete sense. It may then be argued that you can do something for those suffering in your own universe, whereas you cannot reach those in a parallel one, and that that is the discriminator. But as the sum of pain (and pleasure) is always the same if all possible scenarios play out in the multiverse, this is not a possible answer. What good you do here hurts somebody else there! And as all human beings are equal it does not matter who takes the pain.

The Singer theory overlooks how attached we are to the principle of emotional proximity. We concentrate most of our efforts to do good on those to whom we are most emotionally attached. Proximate love is the strongest driver of altruism. Cynics may suggest that, in the final analysis, this is conditioned by our evolutionary interest in passing on our genes, but there is surely more to it than that. We want those we care about to be happy, even if this means that other versions of them in parallel universes, per definition, become unhappier as a result. Altruism can, in a sense, become selfish. We want our universe to be happy, because this is the universe we inhabit and for which we feel directly responsible. Even the remotest human being in our universe can be argued to be closer to us than any human being in a parallel universe. And that is because every being in our universe has a vast number of replicas in other universes. Hence the being in your own universe is, of course, closer emotionally than the replica in another. The *you* you is the 'original' from your point of view, and so are all other beings in the universe you inhabit. You as you want to attract the best possible destiny for your universe of all the countless destinies of all the countless universes. You as you want to attract the best possible destinies for the beings of your universe out of all the countless destinies that will ultimate play out with the countless versions of each being!

The consequence of all this is that it is important for you to fashion the best possible political solutions for the multitude of questions facing your universe. The possibility of earthly immortality will become a crucial political issue, as well as a personal one, if you are sharing the universe with me, and it is in your and my interest to find the best possible solution for our universe – chosen from all the solutions that anyway will come to pass in the countless parallel universes.

Chapter 12
Identity and Infinity

It goes with the territory of radical multiverse theory that an immense number of yous will be spawned every time you make a choice, but also that a large number of yous will arise whenever others make choices and new parallel universes result. Those yous will be even more 'identical' to you than the parallel yous coming about as a result of your own choices, because those yous might at the time of the birth of the new parallel universe be entirely unaffected by the choice that gave rise to their existence. The choice of a Paraguayan will give rise to a new you even if for almost all intents and purposes that you will be unaffected by the choice because that you lives in Sweden.

But also it does not appear to be too much of a stretch to suggest that in the infinite number of universes that may be the consequence of radical multiverse theory an infinite number of identical parallel universes might also exist – and logically that would mean that each different parallel universe would be accompanied by an infinite number of universes identical to it. And an infinite number of absolutely identical yous would hence exist. Such a situation may not be explainable by current quantum theory, even in its most extreme forms, but conceptually it seems possible that if something can be in several states at the same time then it may also be able to be in the same state in several parallel instances. Infinity might not be infinite if sameness is not a part!

That an endless number of identical universes and identical yous might exist would put a twist on the statement that each of us is unique. Despite absolute identicity, the statement is, however, still true from an identity perspective, since identical yous do not put into question the lived fact that there is a you inhabited by you. Oscar Wilde's bon mot: 'Be yourself, everybody else is already taken', may be relativised, but does not lose relevance. Each you is, of course, its own *you* you, unique in a lived 'reality', even if endless numbers of identical yous exist, each with their own claim to being a unique you. The tension-laden truth may be that identicality might not contradict unique identity. Several identical things are still 'several' and, in identity terms, do not conflate even if they remain forever identical.

P. Hulsroj, *What If We Don't Die?*, Springer Praxis Books,
DOI 10.1007/978-3-319-19093-8_12

Finding personal meaning and significance in an ever expanding universe was already difficult when you assumed that there was only one you and one universe. It became harder with radical multiverse theory in its classical form. With an infinite number of identical yous spread across an infinite number of parallel universes it would appear that each you would, in effect, become completely insignificant. Whether a given you lives or dies is of very marginal interest, it could be argued. But that is wrong. First of all, if the *you* you dies all of the identical yous also die, as your destiny is an automatic propagation factor of immense consequence. Yet, it is also true that if you do not die some non-identical you will, alongside all the identical versions of your non-identical you. So, in a sense, that is a zero-sum game, as addressed in the previous chapter. This realisation, however, only points us to the continued relevance of the *you* you. All those other yous may be disconcerting and overwhelming, bringing, perhaps, an even stronger inducement to find refuge in the *you* you. In the unimaginable immensity of limitless universes and replicas, what becomes of singular importance is the *you* you and the loved ones of the *you* you, all sharing the same universe. And for the *you* you death remains a highly unappetizing reality, which the *you* you may, or may not, want to counter with immortality!

In the church of Hornbaek, Denmark, there is an inscription 'That we may live, some must be ready to die'. In a possible reality where all possible scenarios will be played out an endless number of times, it will be true that another you will have to die, if the *you* you decides for immortality. This, however, does not make the choice of the *you* you for immortality immoral. In a zero sum game of life and death, there is no better reason for you to die for another you, than for another you to die for you. All yous have to look out for themselves and their loved ones as their first priority! And every time you survive you will be flanked by an infinite number of yous having equally survived! Thus you are still unique, but, truth be told, no experience you will ever have will be uniquely and solely yours!

Chapter 13
God, Spacetime and the Mathematical Universe

In the Sanctus of Schubert's German Mass God is described as Er der nie begonnen, Er der immer war, Ewig ist und waltet, Sein wird immerdar (He who has no beginning, He who always was, reigns and is eternal, being forever there). This description, however, fits not only the traditional image of God, but could equally well apply to atemporality, spacetime and, in fact, to the mathematical universe propagated by Max Tegmark.

Still, those brought up in the Abrahamic religious traditions tend to understand God as a divine being, as a personalized deity. This understanding has been much helped by Michelangelo and so much iconography showing God as a wise old man with a long white beard. In much of the talk on the beyond there is the same tendency: meeting God face to face, the Lord Jesus sitteth on the right hand of God. And, of course, according to the Bible Man was created in God's image, and the just and the unjust shall be resurrected to be judged 'at the end of time'. Still, there is also the prohibition in the Second Commandment: Thou shalt not make unto thee any graven image or any likeness of any thing that is in heaven…., a prohibition which in Judaism and Islam, of course, is taken very seriously. Although this prohibition was hardly intended to avoid that God is understood as a personalized deity, the effect could to some extent have been that. Or at least respect of the prohibition could have avoided that we would be so overwhelmingly influenced by the beautiful imagery of God, created by most persuasive artists. Their influence infuses Western religion with a need to believe that somebody whose image we share is eternal and that our own immortality equally involves a Gestalt similar to our earthly one, or at least a highly individualized one. Unwittingly we conclude from god to god being person-like.

Of course, in Buddhism[1] it is different. God has been abandoned and instead there are powers and forces that determine the here and the beyond. The fact that Buddhism is non-personalised is surely one of the reasons why it has a strong appeal

[1] Particularly in the Indian version, where also you do not have god-like figures like the Bodhisattvas.

P. Hulsroj, *What If We Don't Die?*, Springer Praxis Books,
DOI 10.1007/978-3-319-19093-8_13

to Westerners with religious affinity and a scientific bent. Er der nie begonnen, Er der immer war fits hand-in-glove also in this context.

Einstein's spacetime creates timelessness by treating time as a fourth dimension; the universe is understood as a permanence where time does not imply change, but is a measure only of position within the timeless 'structure'. Spacetime might possibly not exclude a personalized god, but does not have god as a necessary element. Yet it might be argued that spacetime, in itself, constitutes forces and powers similar to those that could be ascribed to a non-personalised god, and that spacetime confers immortality even on each ingredient within the 'structure', including you and all the other possible yous. Buddhism has de-personalised one branch of religion; spacetime goes a step further by addressing essentially the same forces as Buddhism in purely scientific terms. The operating assumption of spacetime is also Er der nie begonnen, Er der immer war, Ewig ist und waltet, Sein wird immerdar!

Max Tegmark's Mathematical Universe theory suggests that everything is non-material; that everything is purely relational and can be explained by reference to mathematical formulas which unbundle the fundamental and simple relations that ultimately govern physical and other being. Thus Schrödinger's wave function might be the non-material foundation for all materiality. Tegmark's thesis might be too extreme, yet it poses again an interesting issue on being and existence and on what is, in the final analysis, non-religious religion.[2] Although Tegmark might not completely agree, the mathematical relations that underpin everything are eternal forces that animate in a similar fashion as the authority ascribed to god in theistic religion. The Schubert bit is as valid for the mathematical universe as it is for a personalised god.

How does a perspective that focuses on powers and forces reconcile with the quest of humans for immortality of an individualized self? In a sense spacetime and the mathematical universe resonates well with much pantheistic thought, but might pose the same problem in terms of the carrying forward of an individualized self with continued consciousness. All your ingredients, or relations that give rise to the you in the mathematical theory, might be eternal, might be immortal, but you as a unique ensemble might not be, except as a permanent specimen in spacetime.

In spacetime, which is compatible with the mathematical universe, your life was and will remain forever in the 'structure' and hence you could perhaps find consolation in the idea that death would open the door for you to navigate back and forth in your life and perhaps in that of others, including the other yous petrified in the spacetime 'structure'. All the talk about wormholes might give a little hope in this respect, yet no mathematically founded theory implies that death is anything but the end of a particular thread in the spacetime 'structure', the end of your string of 'observer

[2] 'In the beginning was the Word, and the Word was with God, and the Word was God' (John 1.1) seems, on the face of it, to suggest an immateriality of god, not so terribly far away from the Tegmark hypothesis that our universe is entirely immaterial, or, in his optic, mathematical!

'In the beginning was hydrogen' might be understood as the first tenet of the physicist's Genesis. 'And hydrogen gave life to Man!' then connects the wonder of human existence to hydrogen as the first atom and the primary building block of humans.

moments'. Self-aware, self-conscious beings are given no particular status in spacetime, and hence death, although overwhelming in meaning for the individual, has no particular significance.

It is debatable whether spacetime as currently conceived can be the final word on this. Without being anthropocentric it is not an exaggeration to suggest that self-consciousness is a remarkable and unique feature, and to suspect that physics theories, which are inherently orientated towards physicality, might have difficulties capturing this. It is not so obvious why physics accept gravity as a fundamental feature, but not spirit. Yes, gravity can be measured and can be proven, and no, spirit might not be measurable by conventional means, yet it can certainly be proven as unambiguously as gravity. You are the living proof! What we do not know then is whether spirit could be an independent element in spacetime, perhaps even a fifth dimension, which by death would be unleashed to roam within the spacetime structure – in a fashion that is currently everybody's guess!

All definitional work includes a question on the level of abstraction and aggregation. And it is true that in the final analysis spirit might just be another feature that can be captured in an equation, just as gravity, but that does not disprove that at some level of aggregation spirit is its own reality, and does not disprove (or prove) that spirit, at its level of aggregation, might have an independent role to play within the spacetime paradigm.

Spacetime as an alternative to a personalized god might be considered imperfect, since god is good and spacetime is value neutral. But is that really so? If spacetime allows for self-conscious immortality in some form or another, or could be deconstructed to do so, it might involve the possibility of slipping in and out of observer moments, your own, those of all yous, or perhaps completely freely. In radical multiverse theory this would allow the immortal you to partake in all possible permutations of human or, perhaps, even non-human life,[3] bringing spacetime somewhat in the direction of rebirth religions, but also meaning that the menu of experiences that could be re-experienced would include as much pain as pleasure. In that logic one would hope that observer moments could be chosen freely (and how does that work in a non-discretionary environment?) since prescribed sequences of observer moments would tend to sometimes imply moments of indescribable pain. Right, it could then be argued, that is then the difference to the personalized good god. But why are we so convinced that a good god will provide more pleasure and less pain in the beyond than in the hither? With pain being the contrast to pleasure, is it not possible that even a good god will continue blending pain with pleasure even in the heavens? There is a certain inconsistency in assuming that existence continues in a personalised form after death – the face-to-face with god – and that pain, which is a human and personal definer, will be eliminated. In Valhalla and on Mount Olympus there were lots of pain and ample supplies of evil!

[3] Perhaps what you spend eternity doing is then living through all the observer moments of all the yous?

Chapter 14
Winston Decides Not to Die

My friend Marco Aliberti recently said: 'I know I will die some time, but, of course, I do not believe it. So no reason to stop smoking'. This, you may argue, is just a simple self-protection mechanism, in the sense that we do not believe in our own deaths, because the full realization of this might render us incapable of living. But perhaps there is more to it than that. In the radical multiverse theory there is a large number of Marcos that will not die, because earthly immortality is possible. Many immortal Marcos will therefore be a reality in their parallel universes. The bad news for the Marco who lives in the universe occupied by the *me* me, is that this Marco will die, because death is an operating principle of the universe occupied by the *me* me.

In the universe of *me* me, Winston Churchill wrote a memoir of his participation in the second Boer War; a truly remarkable memoir, because the impression is given that Sir Winston never really contemplated that he could die, although he saw his friends fall left, right and centre. To the *me* me this appears completely delusional. Yet, in a very roundabout fashion perhaps Sir Winston was right, that he could not die in the universe we occupy together – because in the confluence of our parallel universes he chose one where he would not die, and therefore could not die. He chose a thread of spacetime where it is hardcoded that Sir Winston will die only at 90.

P. Hulsroj, *What If We Don't Die?*, Springer Praxis Books,
DOI 10.1007/978-3-319-19093-8_14

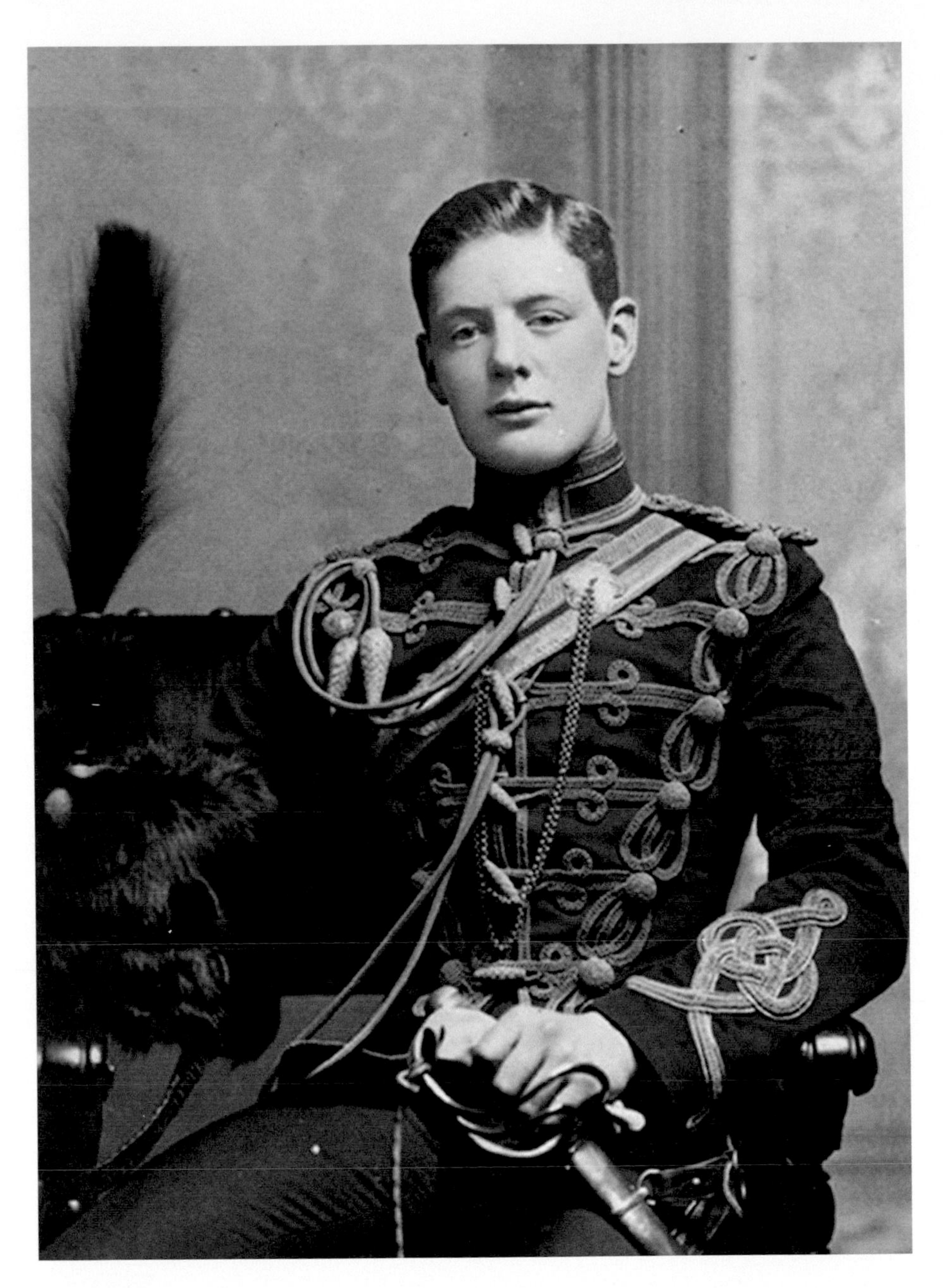

I cannot die!

What is really at issue here is to which extent every you and every me can choose the path we take within the innumerable superpositions and resulting parallel universes; whether the Sir Winston who wrote the book was right that he could not die, because his version of Sir Winston had chosen to occupy a parallel universe where South Africa would not be his place of death – and, further, that this Sir Winston had a premonition of the universe he was steering through. Perhaps the sort of courage that is easily mistaken for lack of imagination is really a highly developed sense of the direction of the thread of spacetime occupied by that version of the person.

Now, discretionary choice, free will, is not easy to reconcile with spacetime as a permanent structure with no start and no end. But perhaps 'free will' is exactly exercised by every you within the alternatives offered by superposition, with the consequence that an alternative you is created that will have as baggage the alternative not chosen by the *you* you. So, the thread of *you* you is a consequence of all your choices. And the consequence of the immense number of choices of the *you* you is that you spawn an immense number of parallel yous, which share everything with you up to the point of choice, and which afterwards exist as different yous, each of which will again face an immense number of choices, giving birth to an immense number of parallel yous to your parallel yous – each being an Abraham of discarded choice replicas and their replicas. The mortality of *me* me is possibly the result of a very bad choice when the *me* me was divided as egg and spermatozoid!

Many people seem to have a somewhat undefined feeling that one avoids bad things if one worries enough about them. And they may feel that when bad things happen they are often unexpected, and perhaps a result of not having had the occasion to worry away those bad events. Who knows if they are not right? Perhaps the worrying is what has steered the worrier onto the desired path – with the consequence that another version of the self will be saddled with the feared event. Similarly, many conjure up images of a desired future, awake or in dreams, and have a feeling that the dream is reachable, but will not be reached. And, of course, the dream is reachable in the sense that it will play out in some parallel universe with another version of that self enjoying the benefits. Perhaps the self with the premonition of failure in reaching the dream is merely realising that this self does not have the strength, the karma if you will, to make the dream reality, does not have the strength or the conviction to steer the self in the desired direction, cannot 'break on through to the other side' as the Doors, as mentioned, would have it. You may have the choice to realise a dream, but does the *you* you have the strength? Perhaps that would have required that the *you* you had made different choices earlier on in the universe the *you* you inhabits.

Philosophers and teenagers alike have long asked whether the world they each live in is made just for them, or whether there is an objective world out there. Surprisingly, the answer might, in fact, be that the world, the universe, you occupy is to a large extent made by and for you. The universe you occupy may be the result of the choices you have made. An enormous number of other yous occupy parallel universes reflecting the choices you did not make. And perhaps somewhat troubling, the ones you love also exist in a multitude of forms, each interacting, or not, with another version of you. The *you* you can only embrace the loved one who has made choices making this version of this person part of the universe of the *you* you. Each time you embrace, you give birth to a parallel universe where another you does not embrace another version of your loved one.

Schopenhauer spoke of the will to life, Nietzsche about the will to power, and humans may surely also have a will to happiness, a will to tragedy, so many other possible wills to something. In front of us there may be an immense number of paths, masks, and our every choice might therefore be a choice between so many different futures. The Sir Winston of the parallel universe shared with the *me* me may have chosen a stencil that enabled him to survive the war. This could not be redone, because the choices we make, even in this theory of greatest freedom, will still be final. However, new choices will make the *you* you follow a further mask chosen among many masks, but based on the status of the mask that brought the *you* you to the choice in the first place. Sir Winston could still have summoned demons, but this Sir Winston knew that he would not. Another Sir Winston knew that he would and he will share another parallel universe with another me. All yous may weave in and out of masks, but no you will be able to retrace the tracks for the part of a mask the you has already trodden.

Less radically, there might not be a free will for each you to choose which part of the spacetime braid that you shall occupy. Nevertheless, the result of the application of a set of laws of nature will have endowed any you with qualities that make predictable how the spacetime thread of that you will look. Perhaps a feature of many threads is that the you who is in residence can foretell how it ends, à la Sir Winston. In fact, if nature makes sure that all permutations on all themes will always play out, then there will be as many yous who cannot foretell their future as there will be yous who can. Your terrible problem is: is the *you* you one or the other? By definition the premonition of at least one you will always be right,[1] but is it the *you* you?

In addition to this highly disconcerting issue there is an even more fundamental one: does not the reality of the *you* you mean that you must have discretionary freedom to choose? Is not exactly choice the meaning of being exactly you? Are not the choices you make as you, the element that makes you what you are – is not choice the whole point of having consciousness?

[1] Which, of course, may then be an illusion, and the veracity only a result of statistical probability.

Chapter 15
God Is and Is Not

God as the creator of the earth and the universe is a central theme of the Abrahamic religions. And that is true even if evolution theory is accepted and crude versions of creationism rejected. God is still the Creator. With this approach the question does not come so easily whether God himself was created, and by which processes. In non-theistic religions like Buddhism, obviously the question does not arise, and the issue becomes the nature of the forces that created the universe. In Greek mythology Chaos, a primeval void, gave birth to Gaia, the mother of the earth, and mother of Cronos, the father of 'the father of Gods', Zeus. In Norse mythology everything, including the gods, came about as a result of the heat of the south meeting the cold of the north in the great void. Odin, allfather of the gods, is himself the son of Buri, a proto-type god. Odin, with his two brothers, killed Ymir, a giant and the first being, and created the world from his body. At which point they also created time!

In Christian theology Christ is descended from the Father, and in this respect there is similarity to the early theistic religions. However, the Father himself is not explained in terms of genesis. In a sense this might point to God as a quality, rather than a personified entity, and could thus be argued to show kinship to Buddhism.

If the divine is a quality, or if God is created by natural forces, in addition to being perhaps the Creator, radical multiverse theory would seem to imply that God can exist in one universe and not in another, and furthermore that if the divine can possess different qualities then different gods can exist in different universes alongside godless universes and universes possessed of the non-theistic spiritual qualities set out in, for instance, Buddhism. Some gods may bestow eternal life on humans in the beyond of their specific universes, others may not. Quantum theory implies that all solutions that are possible (in a very wide interpretation of the possible) will be reality in one or more parallel universes. Thus also there may be many identical or very similar gods in the numerous parallel universes, just like there may be so many yous! Yet, if god is created it is not obvious that there can be one god for all universes, since the parallelism would seem to militate against an overarching force. Just as you cannot transcend your universe to seek communion with your other yous in

P. Hulsroj, *What If We Don't Die?*, Springer Praxis Books,
DOI 10.1007/978-3-319-19093-8_15

other universes, so a god might not be able to eclipse all universes unless god is a quality that can permeate all possible universes – unless god is one of the absolutes of the laws of nature, rather than a personalized deity.

By the same token all possible forms of evil and the devil will exist in one universe or the other. All possible combinations of good and evil, gods and devils, will play out, and, to the extent different forms of divinity and evil are not mutually exclusive, some universes will have several gods and several devils. Thor and Zeus together fighting Typhoon and Loke might be the mythological illustration! And, if we believe the Norse, there will be universes with time and universes without!

So if two qualities can co-exist in some universe they will. If, according to absolute laws of nature, two qualities cannot coexist in the same universe they will then co-exist in separate parallel universes.

Perhaps quantum mechanics is the great uniter of all that seems contradictory! Perhaps quantum mechanics is the ultimate embodiment of the first principle of epistemology: that reality is almost always wilder and more complex than we imagine!

Chapter 16
Linear and Non-linear Logic

The progress of Western philosophy has largely been based on application of linear logic. This is true for Hegel and Kant and true for Plato, Aristotle and most of the philosophers travelling the road of Hellenic philosophy. Cartesian logic prides itself on its linearity: you build a chain of analytically connected arguments, each of which you 'know' to be true, one 'truth' being the stepping stone to the next. Linear logic appears to a Western mind to be inseparable from the logic we find in the physical sciences. Notwithstanding Hume,[1] the cause and effect paradigm of analytical logic seems to us to be indisputable and leading to internally coherent 'truths' eliminating the possibility of the veracity of competing 'truths'. In fact, a main objective of our philosophy is to find the 'one truth' or 'the correct answer'.

Asian philosophy, to the extent it can be generalized, is far less obsessed with eradicating internal contradictions in concepts or logic. Absolute truth is questioned even more than in Western philosophy – is considered less relevant. 'The one correct answer' is less a concern than capturing the nuances of the question and the nuances of the answer, even if these nuances of question and answer might be non-absolute and sometimes appear contradictory or as expressing parallel 'truths'. In jurisprudential thought Amartya Sen, no slough in linear logic, has put a considerable question mark against the usefulness of trying to define ideal justice, simply because several contradictory ideals can be defined, each being rational and irrefutable on its own premises.[2] Sen thus argues for incremental justice, since some situations are so manifestly unjust that no rational philosophical approach can put this into doubt. Although Sen draws on a wide variety of sources from many corners of the world, clearly quite a lot of inspiration for his ideas on justice is drawn from Indian philosophy with its greater ease in dealing with opposites without necessarily having to resolve inherent contradictions.

---

[1] Who himself, of course, applies linear logic in the most rigorous fashion, even when challenging cause and effect outside logic.

[2] The Idea of Justice. Interestingly in the context of the current book and for a professed atheist, Professor Sen's first name, Amartya, means immortal in Bengali.

P. Hulsroj, *What If We Don't Die?*, Springer Praxis Books,
DOI 10.1007/978-3-319-19093-8_16

Overall, one could thus have the impression that Asian philosophy is less aligned with the physical sciences than Western philosophy and this seemed convincingly true until the advent of quantum mechanics with its normative anarchy. However, quantum mechanics confront humankind with dilemmas, such as the possibility of being in several places at the same time, and ultimately with the uneasy realization that at the quantum level, linear logic cannot apply. Computer science has embraced this breath of fresh scientific air with enthusiasm and quantum computers are on their way, utilizing inter alia the ability to be in several physical states at the same time. Western philosophy has, by and large, not followed suit. Niels Bohr famously said: 'If quantum mechanics hasn't profoundly shocked you, you haven't understood it yet'. It would appear that Western philosophy, so indebted to physical sciences, indeed 'hasn't understood it yet', because it has certainly not been shocked into philosophical action to embrace the most profound consequences of non-linear reasoning.

After more than 2000 years of the application of linear logic, Western philosophy might have to accept that the paradoxical, non-absolute, nuanced, parallel 'truth' paradigm of Asian thought has a role to play alongside the rigorous linear logic of Kant and Hegel. When Niels Bohr designed his coat of arms for the Danish Order of the Elephant he put as his motto: contraria sunt complementa. We have a long way to go in philosophy before we have really come to grips with how opposites are complementary!

CONTRARIA SUNT COMPLEMENTA

Generally speaking, this book discusses immortality in the classical linear logic tradition, but it is a useful caveat to remember that conclusions must be informed by the lessons of quantum physics – or, you may argue, by the suggested first principle of epistemology: that the reality is almost always wilder and more complex than we imagine. Indeed, parallel realities might abound and truth might be inherently contradictory. That, however, is not a reason not to reflect on truth and reality and on how we as a human society will act within the realities we understand and can influence, particularly because the basic assumption of this book is that the interest in immortality is in a continuation of personal identity. With that starting point we focus on sensory experience which, even taking account of Freud and Jung, is a unitary perspective: what we feel we are! Almost cogito ergo sum!

Chapter 17
What Is Immortality?

The concept of immortality raises many more issues than just whether eternity is poetic, atemporal or sempiternal.

In fact, it even raises a question of perspective. Previous chapters discussed whether a continuity of consciousness is the mark of eternal life; the perspective in that case being exclusively that of the aspirant to such life. Yet, identity can also be understood as the relationship to the Other, as propagated by Emmanuel Lévinas, borrowing heavily from Hegel via Heidegger, and stated so nicely by Einstein: Only a life lived for others is worth living. The essence of life and of identity can be argued to be the relationship to others, although this would be disputed by many a saint on a column and by Jean-Paul Sartre, for whom Hell was the others. And it is certainly relativised by Kierkegaard's, 'Once you label me you negate me'. This admonition is worth keeping in mind: an externally defined identity will never match the internally defined self in terms of comprehensiveness, depth, complexity, shifting nature, or nuance.[1]

Defining identity through others would, however, have a fundamental impact on the question of the immortality of that externally defined identity. 'Tell me who your friends are, and I shall tell you who you are' the popular saying goes, and although the inversion 'Tell me who your enemies are, and I shall tell you who you are' is more interesting, both statements point to the externality of identity. Almost everyone of a certain age has experienced meeting somebody seemingly knowing you very well, although you have no clue who they are. In those situations it is hard not to rejoice in the friendship and assume the mantle of a history you no longer know. Your scoreboard

[1] The cult of heroism, which humankind has practiced continuously since its early days, think Homer, is an epitome of externalism. Inspiring heroism might have been useful to allow humankind to face the many challenges to its survival, yet suppressing common sense, fear and empathy carry also great risks as centuries of pointless warfare has demonstrated. Rupert Brooke took the externalisation of the self to frightening heights in his poem The Soldier. The physical remains of the soldier buried in a foreign field becomes a part of England.

P. Hulsroj, *What If We Don't Die?*, Springer Praxis Books,
DOI 10.1007/978-3-319-19093-8_17

is blank and yet you fit hand-in-glove. The Other has defined you![2] Studies seem to show that the development of personality is to a large extent a reaction to the influence of family, friends and foes, some going so far as to suggest that parents influence the development of the child's personality much less than friends and foes.[3] Conversely, evolution theory claims that off-spring and kin are the main conditioners of behaviour and thus of identity, that they are the significant Other, whilst various humanist theorists, such as Lévinas, speak more broadly of the loved ones or a circle of friends. Although the humanist and evolutionary points of view are not entirely at odds, a choice between their underlying concepts leads to fundamentally different results when it comes to the question of how to understand immortality. The passing on of genes as one kind of possible immortality will be addressed later, whereas immortality as an externally defined identity in a broader sense will be reflected upon immediately.

Much depends upon form when discussing externally defined identity and externally defined immortality. The portrait in the gallery of the ancestral home will keep your identity alive, narrowly and for a while. The beauty of Shakespeare's words gives him a rich life beyond the dust of his physical being, although, in truth, we know little about the man except what we surmise from his work. A very indirect immortality, indeed, although one that Sartre chose as his escape from extinction and Horace celebrated with 'Not all of me will die; part of me will escape the goddess of death'.[4] In our fame obsessed culture even the most bizarre moment of celebrity seems to give meaning not only to life, but beyond, for the ones so blessed.

'I'm gonna make it to heaven,
Light up the sky like a flame,
I'm gonna live forever,
Baby, remember my name,'
as the eponymous song from the movie Fame proclaimed.

'Eternally owned is but what's lost'[5] is a more poetic expression of the possible immortality of memory. That good deeds, and, alas, bad ones too, survive in the lives of family and friends is poignant testimony to our complex humanity and our balance sheets of interpersonal achievement, but is it immortality? If it is, it is a Buddhist/

[2] There is a related beauty in Karen Blixen's description of how some of the Africans she met dealt with affection: A white man who wanted to say a pretty thing to you would write: 'I can never forget you'. The African says: 'We do not think of you, that you can ever forget us' Out of Africa, 85.

[3] The labeling theory of George Herbert Mead is all about how external factors influence the development of the self.

[4] And in Sonnet 18 Shakespeare himself chimes in on how his poetry is not only eternal, but bestows immortality:

But thy eternal summer shall not fade.
Nor lose possession of that fair thou ow'st;
Nor shall Death brag thou wander'st in his shade,
When in eternal lines to time thou grow'st:
So long as men can breathe, or eyes can see,
So long lives this, and this gives life to Thee.

[5] Henrik Ibsen, Brand, Act IV.

Hindi echo, our scorecards being carried forward, if not in rebirth then in the continued lives of those we touched. 'Unable are the loved to die, for love is immortality'.[6]

Eternal life is, of course, not just a very long time but without end, and in this respect even our best deeds and worst crimes will fade and finally disappear, except in the sense of our actions predicating the future as was so nicely described in Ray Bradbury's A Sound of Thunder, one butterfly of the past killed and the destiny of the world was changed. Our actions by definition live forever there Ibsen is certainly right but remembrance goes. Henry VIII is part of the common, cultured consciousness, the tall tales and the portraits, but in 10,000 years will he be any different from some Egyptian pharaoh, sun of his time and now not even dimly remembered? Will Shakespeare's dramas be even remembered if humanity survives for another million years? The most beautiful book title, A la recherche du temps perdu,[7] is right – time can be lost, and, in truth, rarely will it be re-found. And when, then only like the insect frozen in death in a piece of amber, its sting no longer even an irritant. As Kierkegaard said, diametrically opposed to Ibsen: 'for it is not worth while to remember that past which cannot become a present'.[8]

Time is a measure of quantity, not quality, and sempiternity is a measureless measure of an unwritten canvas. It can be argued that being fixated on immortality confuses the discussion by not looking at quality. Perhaps time is not as relevant as experience, so if we seek the unbounded for ourselves then perhaps unbounded experience, rich experience, should be our ambition. Apart from mysticism this moves us again to atemporality, all things at one time, and for this earthly immortality provides no path; if a path exists at all, death is the door. Thus, the dilemma remains how temporal beings can partake in atemporal experience, when the two qualities seem to be fundamentally incompatible.

The quality of life versus quantity of life will remain the schism in the analysis of earthly immortality. And this schism is partly informed also by 'fame' or achievement considerations. When Ingmar Bergmann died Woody Allen said that he thought that Bergmann would have been ready to forego a few of his masterpieces for a few more years of life. Although the works of Bergmann, indeed, show a great horror of death it is debatable that Allen was right, simply because his view ignores what defines us – even what we are evolutionarily conditioned to let us be defined by. His view ignored the evidence of the many bright stars who burned out so quickly in the pursuit of literary immortality; Shelley, Byron, Keats put themselves on the high-wire, and fell, and many before and after them did the same without achieving the pay-off that those young poets eventually did. Bergmann surely wanted more years, but it must be doubted that he would have opted for more years at the expense of what so fundamentally defined him, not just in fame but in terms of the quality of his earthly experience.

---

[6] Emily Dickinson,

[7] Marcel Proust.

[8] Fear and Trembling, 22 (Aristeus Books). Samuel Scheffler seeks in Death & the Afterlife to show that our personal value systems are so profoundly influenced by the importance of continuation of the human species that our post-death 'afterlife' at the bosom of humanity can outweigh our interest in personal survival. Yet a collective afterlife is a far cry from the personal afterlife which so many hanker for.

The way to gain immortality?
(Credit: Joris Thys)

Even if it is not assumed that a cosmic law of balance means that every human has only a certain amount of quality life experience that can be tapped, a million years of earthly life for an individual quickly takes you to Nietzsche's eternal repetition,[9] and the question of how one can fill a life that will never take an end. Susan Ertz has put it pithily: 'Millions long for immortality who don't know what to do with themselves on a rainy Sunday afternoon'.[10] Sex is for many the strongest impulse, and yet even in this field endless repetition diminishes the attraction.

We talk about immortal love, yet what is love's future when we are called upon to prove exactly the immortality. Can any partner be interesting enough for a life together for a million years? No matter that the partner also grows and experiences over the million years, it is a high requirement to remain an interesting object of emotional exploration for such a time. Almost all relationships of longer duration contain an element of 'taking for granted'. A time horizon without horizon will move 'taking for granted' to emotional imprisonment and ultimately to emotional rebellion. No sensual love will survive eternity. Atemporal sensual love is conceptually possible, sempiternal is not. Sempiternal sensual love is frightening, in fact – a boundless desert you will be forced to walk through without water, ever thirsty, yet never dying from thirst.

In Harry Potter professor Dumbledore convinces Nicolas Flamel and his wife that after more than 600 years of life enough is enough. And the idea of 'enough' is the rub here. Once you have accepted the reality of 'enough' then logically it applies to all quantities with the same characteristics. This realisation might not challenge the concept of eternity directly, but it negates its desirability. Life, you might say, is not the same thing as marmalade and hence 'enough' might not apply. But are marmalade and life not, in this sense, the same thing?

Still, let us not rush to conclusions; there are more kinds of immortality than meet the eye! One of the most fascinating ones leans heavily on Jung's idea of the collective subconscious and can be argued to be the step beyond evolutionary group selection theory. An experiment performed by the University of Pennsylvania has demonstrated that electrically conditioned telepathy can transfer experience from one rat to another far removed. Electric impulses allowed one rat to steer another based on what the first one had experienced. If a broadening of this technology were possible a neural network could be imagined where a number of persons were able to share in the experiences harvested by each individual in the network. And when one member of the network died the experiences of that individual, as part of what would ultimately be a collective personality, would live on in the group. In some respects this is just a more radical permutation of the idea that humans will be able to transfer their consciousness to computers; another kind of possible immortality, but one that possibly could also be scaled up, since there is no particular reason why it should not be possible to join together different persons living machine lives.

[9] See also Bernard Williams, The Makropulos Case: Reflections on the Tedium of Immortality, in Problems of the Self.

[10] Similarly Anatole France: The average man, who does not know what to do with his life, wants another one which will last forever.

A collective consciousness hosted by a machine! And actually there is no reason why this collective consciousness could not be enhanced with the cool calculatory power of the machines themselves. The borders between individuals and between individuals and machine would become almost imperceptible and questions of identity, individuality and humanity would change into questions on communal identity and humanoid character.

The first step on that path might be mental pace-makers which might be almost within reach. The normal pace maker keeps up the beat of the heart, a mental one might speed up the slow or the slowing mind – the artificial intelligence of a small computer possibly linked to our organic one, allowing our brains to command the power of the machine whenever our natural ability poses a barrier. From there it is only one more step for the brain to be able to command access to the collective knowledge of the internet, and from there another small step to collective consciousness. Google glasses, in a sense, show the way.

The idea of telepathically networked collective consciousness is perhaps less frightening than machine-based collective consciousness with humanoid extras. At least there is a residual physical identity that gives the comfort of a recognisable humanity. Yet, even there we move to a plane never foreseen by the Marxist celebration of the collective, since the emotional distinction of the individual melts away when all emotion and experience are shared. With collective consciousness sexual identity will disappear as well, and procreation, to the extent desired or allowed, would become hermaphroditically difficult when you would be on both sides of the sexual act or, indeed, on both sides of a mass of sexual acts performed by different bodies but amalgamated into a common collective consciousness. Gaining a God-like understanding of all emotions and experiences might make the human species immensely intelligent and perfectly empathic, might be just another evolutionary step from single cell organisms to all-encompassing identity integration, but where does that integration end: with all humans as one, as all higher order living beings becoming one, or all living beings gaining a single identity dissociated from physical existence of any one constituent part – and therefore also living forever -more? With the brain of any one individual very limited in the face of the richness of collective experience will humanity become a distributed sensory system akin to distributed computer systems and with the same kind of selectivity in terms of data preserved, meaning that all members of the network will contribute to the sum of experience flowing though the system, but without assurance that any element of own experience will be retained by the collectivity for any length of time? This kind of immortality has nothing to do with the earthly immortality discussed earlier, it is more like the immortality that can be imagined when Western religion talks of 'going home to live in God', as discussed above in Chap. 8. The collective transcends the individual and thereby becomes thc new, perhaps immortal, individual. Technically enabled pantheism in a very earthly sense.

As discussed by Locke identity is always definitional: Is a leaf a separate identity although being part of the identity of a tree? Is a tree a separate identity when part of a wood? – a scaling up that can lead to Gaia theories and the universe being the only identity (in the absence of proof of the multiverse). What networked telepathy adds is only a much tighter identification or empathy between human elements. From the perspective of the individual human being this is a paradigm shift, but is it a shift from any other perspective?

Machine-facilitated collective consciousness is barely ontologically understandable. If physicality becomes entirely irrelevant and any sensory response machine generated, if all reasoning is determined by the logic of 0s and 1s, did we not create a global identity that might be immortal but entirely impersonal and hence not worth having? Would machine-facilitated collective consciousness not lead to the ultimate human Selbstaufgabe, where the starting point might be an amalgamation of individual experiences but where these experiences lose shape in the same manner as a drop of water does when it becomes part of the sea? If self-preservation is our ultimate goal, machine-facilitated collective consciousness is the anti-thesis.

The same is not true for machine-based individual consciousness. If such a transfer becomes possible we might lose part of what we until now have understood as our humanity, but we might also be machine-enhanced in terms of reasoning, we might gain a chance at eternal life, and we might make it possible to accommodate many more new generations, as machine-based life might demand fewer resources. But exactly this question of new generations highlights another ontological dilemma, and that is: what is a new generation? If newborns are transferred to machine life the machine must capture the whole miracle of the DNA roadmap in order to let the new being develop as it would have physically – or do we wait until maturity has been reached, thus combining physical and machine based life; the consequence being that any physical clock is stopped at the time of machine transfer. Or, even more radically, do we replicate our entire lives, development, aging and death, only to live on after death in a machine-created Paradise, with restored full cognitive powers and full recall of the cycle of birth to death to rebirth? A Buddhist/Hindi paradigm, but frozen in one rebirth – or in more, if the first time around was not good fun, we stopping the rebirth sequence only when we have tried out many different lives, and gained a kind of Nirvana, with full recollection of all our lives, and a conclusion that no further new life is worthwhile – at which point the individual identity will have achieved a kind of atemporality, the clock stopped for all things but continued individual reasoning in infinity. Not surprisingly, this does not sound much more attractive than collective consciousness!!

Machine-based continued consciousness obviously opens all manner of questions about sensory ability, mobility, and more fundamental mind/body issues (does only mind define ‘soul’, M. Descartes?). The choice of a machine-based life is in many ways counter-intuitive and at odds with evolutionary conditioning, seems to come close to choosing to become a quadriplegic. But ontology is challenged in an even more fundamental way, as a transfer of identity would seem to allow for duplication of personality, for radical twinning and ultimately for cloning-based procreation.[11]

---

[11] Derek Parfit in Reasons and Persons has given interesting, actual, examples of divisions of ‘the self’, and introduces extensive analysis. His purpose is definitional in relation to the concept of identity, however. In his ultimate rejection of definition of identity, in favour of pure factual

The new you?
(Copyright: University Vienna)

The point is that a transfer of identity to a machine would seem to be less a transfer than a copying. What would happen to the remaining physical substance? God would not suddenly have withdrawn his breath of life; left on the floor would not suddenly be a lifeless mass of flesh with no mind. No, if such a thing becomes possible then you as a physical being would continue, and your mind would continue storing your experience and your capability. What you would have done would have been to copy your identity onto another medium, which thereafter might assume independent existence, but will carry the identical life baggage as you from the past. Derek Parfit is a master of thinking up such examples.[12] Logically there is no reason why this process could not be repeated endlessly- through cloning each individual could become the stirps of endless replicas, a possible megalomaniac nightmare based on a modern version of the Bokanovsky Groups of Brave New World! And, of course, how the clones develop will depend of the sensory apparatus with which they are equipped. If all were hooked into the same source of pleasure and pain, hunger and nausea, an octopus in reverse with only one arm but an endless number of heads, then all twins would develop identically, assuming that the hosting machines were also identical, which, of course, over eternity might be difficult to achieve, as one may have a power cut, another not. But the basic concern is this: what would be the point of huge numbers of identical identities? Also this would ultimately break the logic of identity. How could anybody care if one was lost or not, and if the machines did not communicate they would not cooperate, every machine would be less than a sand-corn – fulfilling no function, having no social interaction. Assume that the computers will communicate then. A central machine would distribute tasks and machines would cooperate to fulfil them; their experience becoming diverse. Then we move towards the machine-based collective consciousness, but be worse off because the holders of collective consciousness would all have the identical starting point in terms of identity.

Like successive saves on a computer it can, of course, also be imagined that a person can choose to upload his or her personality at different times, thereby creating different copies of his or her identity as a function of time: 'Oh, I feel particularly good today, let me upload!' That added choice might increase the attractiveness of this kind of procreation, but will ultimately not resolve one of the fundamental problems, which is that by stopping genetic mutation you create a world frozen in genetic time. Over eternity quite a boring perspective!

The attraction of this changes considerably if ultimately machine-based life can become akin to life in a bottle. What if you upload your identity and deprive it of new impressions? You create almost local atemporality, storing your life so far, and you give it soul by creating a search mechanism that will allow this life to be lived over and over again. A modern day Nietzscheian 'endless repetition'. Intuitively, of course, you feel claustrophobic, yet if you upload your life every month you replicate your soul, so far, every month. Eternal life not once but many times

description, Parfit takes the philosopher's path. He does not address the consequences for society of possible extensive replication of the self.

[12] Op. cit.

over – what more can you wish for? Parallel universes might or might not exist, but you will have ensured that on the personal level they will! The existential question will be whether soul, as opposed to consciousness, can be replicated. Until now we thought that 'soul', whatever this means, is a unique identifier: indivisible! Lord Voldemort of Harry Potter fame divided his soul in seven in order to become as indestructible as possible, but his consciousness stayed one, so that is a far cry from dividing into many separate consciousnesses with the same soul replicated for each consciousness. Replicating souls endlessly is in a sense a hyper-charged version of David Hume's assumption of personality renewal every single moment. With Hume at least there is a sequencing of renewed personalities, in machine duplication there is both parallelism and sequencing: Hume squared! Or, in a sense, technically enabled multiverses, with endless numbers of yous being created, but with the *you* you nevertheless being a unique and perishable soul!

When considering these wild possibilities mere physical immortality seems almost parvenu. Yet also physical immortality raises issues of how elements of our being can be separated out and put in new contexts. The simplest example might be cloning. Is the simplest kind of immortality that you perpetually clone yourself – that every time your physical clock is running out you create a new self? Without your consciousness it is true, but still a higher degree of immortality than the one where you have to mix your genes with those of a partner. And, just like for machine-based cloning, you can be more ambitious and become a new Abraham and populate the Earth not just with your off-spring but with your entire likenesses. Such an achievement is the ultimate in terms of complying with evolutionary conditioning: you pass on your genes in unchanged combination. The question arises, however, whether such broad-based replication does not introduce a hitherto unknown kind of genetic competition. Assuming that genes always compete not only at the level of individual genes as in classical evolutionary theory, but as a body of combined genes, there would seem to be a strong impetus for one such body to eliminate another identical one in order to facilitate the continued existence not only of identical sets of genes, but of exactly that set of genes. Darwinism with an even sharper edge! As no partner will be necessary for procreation, genetic variability will provide no protection against genetic imperialism, everybody might aspire to become the new Abraham! It might become not just armies of clones against armies of other clones, but clone against identical clone in an even more deadly competition than until now, simply because the Other will no longer be of relevance for your 'immortality' goals. Our current need to find partners for procreation, and partners with good genes, has been a fundamental humanising factor. Despite so much supremacist thinking, genetic variation has been the civiliser; a lesson which is tragically forgotten with every ethnic cleansing, with every raised barrier to immigration.

Physical immortality also raises many of the same questions as did machine-based immortality. Will our future genetic mastery allow us, even more than Dorian Gray, to time our aging as we wish? 'Oh, I will take another year of being 19', whilst my younger brother has decided to race to 31 – in the process becoming my elder brother, physically at least. And, even more perplexing, will we be able to go back and forth in age, at one time being senior to our parents, at one time being a baby to our middle aged 'younger' sibling. The social variability would be tremendous, with one truly constant in our current social relations, relative physical age, becoming a complete inconstant. Assuming the ability to be able to stop the clock, but not go back, we would be faced with a fundamental dilemma, when to stop the clock definitively, when to freeze our age? Is the prefect age 19 or 61? Do you go slowly, slowly to 61 and then choose to stay 61 forever? But is not 19 the perfect physical age and since our consciousness continues is that not all you need to choose? A world of only 19 year old physical specimens, but some having lived 600 years and some a mere 19? The adoration of youth will be physical only; the mental energy and boundless optimism will no longer be signalled by physical appearance. There is, in fact, even a mind/body issue in this as well. If you live in the perfect body forever you will not experience the physical aging that is such a defining feature of current human life. And a 19 year old body which is not preparing for the battle to death with age is not the kind of 19 year old body we know now. What we live and understand is, to a large extent, defined by contrasts. Eliminating death will redefine life fundamentally, as its contrast will have disappeared. The price of immortality is that the flowers of life will shine less brightly. Of course, we will still see animals and plants die, but when death does not apply to us the contrast to our immortal lives will not be so clearly appreciated. Why enjoy a swim when you know that an endless string will follow – the horrors of Nietzsche's endless repetition become palpable. First generation immortals, those born to die but saved, will enjoy immortality more than those born to immortality, but both will probably end up jumping off the cliff after a life less luminous, less informed by contrast, but much longer than that of current mortals.

When talking about immortality it was just discussed how the assistance of machines could put the traditional concept of humanity into question. Genetics might test this even further. In the quest for immortality one step might be to introduce non-human genes into the human make up. When this is done to fight a specific disease this might not create existential issues, but when longevity or other characterising features become dependent on animal genes we are on the path to creating chimeras and to the total blurring of the human and the non-human. Essential ethics topics will arise including who can be legitimately killed for food and who is protected as part of the human species; who will be entitled to immortality?

Will any being carrying human DNA be protected, and how would this affect the opposite route to the one just mentioned, namely when we inject human DNA into animals or even plants? Immortality seems from afar such an attractive concept, but when getting closer we see that it becomes difficult to determine whom we serve and that the means of getting there make this dilemma more and more pronounced. When machine interacts with human and human with animal and plant, thus creating hybrid identities, all our most fundamental beliefs and understandings of self are put into doubt. It is easy to assume that progress is unstoppable and that all moral issues will be resolved by themselves along the way, but the reality is that the human future will be replete with moral choices that will define what we, from a human and global perspective, will allow as pursuable progress. In the past 'doable' defined progress, but in the future we will increasingly have to define up front whether doable will, indeed, bring progress. It would be comforting to think that we are preparing ourselves as a global community to face such a fundamentally different challenge!

In a sense, the final frontier in the quest for immortality is the creation of synthetic genes. This might appear beyond outlandish, because it would imply an ability for humans to create artificial life ingredients, but it is probably not. What nature was able to create humans will normally be able ultimately to recreate, including new forms of the building blocks of life. It might take time, but remember that our time-horizon is eternity! Where would such an increased mastery of life leave our humanity, however? Are we human if we consist of 50 % artificial material? It is true that we have artificial limbs and their use does not affect our humanity. But what when our brain cells are half synthetic, when our skin has become coolly synthetic, our ability to bear pain much greater, and our responsiveness to pleasure immeasurable? Do we move beyond Spiderman to not being Man at all? Or is synthetic life just another version of machine-based life?

Where all these questions on enhanced, synthetic and machine-based life come together is perhaps in the possibility of a replacement virtual world. If we do not need ourselves as we know us, do we need the world as we know it? Is not the ideal world one where we have improved creation's work by making the Earth (or elsewhere) the mere host of our physically much reduced, but immortal beings, but without the Earth or other beings having to provide any kind of sensory input, as all will be provided by our virtual worlds? A human universe populated by a wealth of parallel dehumanised personal universes – existence where we have abandoned humankind's traditional exploration of the external world for all-embracing introspection. So taken to its ultimate consequence, progress might take us to a place where the Other is entirely irrelevant and where human interaction is in fact entirely imaginary. Poor John Dunne: 'No Man is an island' turned into 'Every Man is an island'! And yet, as you will never know the difference, is there a difference?

The Bible speaks of how the meek shall inherit the Earth and the poor in spirit the kingdom of heaven. Perhaps there is an admonition here that pursuing the route of the apple of knowledge will lead to perdition, and that hence we must remain meek, poor in spirit, even in the face of our human potential of invention; that we must abstain from some human inventions in order to stay human! Yet, the Biblical tale of Lot's wife cannot make us optimistic. The untameable curiosity of Lot's wife turned her into a pillar of salt!

Chapter 18
Is Anything Forever Existing?

As shown, there may be many ways of living on beyond our current physicality. But even if death can be eliminated from the equation in the first instance does this mean that immortality is truly possible – or only the longest life? Some will argue that nothing is truly of permanence, except possibly god, and even that can be debated, as done later on.

Staying in the realm of nature it can be asked whether nature knows permanence, immortality, indestructability. What can be observed is that everything gets worn or dissolves, but this, of course, does not imply that there is no permanence, since natural science is premised on the assumption that nothing vanishes, it only changes form. In this logic the constituents, whatever they be, are eternal.

Yet, even the natural sciences must perhaps distinguish between the eternal and the seemingly eternal. Our Universe may have started with the Big Bang through which elementary particles arose, allowing hydrogen and helium to be formed ultimately, the fundamental building blocks of the Universe. And there might well be protons which have remained unchanged since their creation just after the Big Bang – as a result having been permanent for almost 14 billion years. But ultimately our Universe might crumble, might experience the Big Crunch, and, even if not, protons might become instable and decay, albeit in a timespan much, much longer than the currently predicted life of the Universe. Yet, this does not necessarily put into question that 'nothing ever became nothing' or its attendant 'nothing came from nothing'. Singularity might have ruled at the time of the Big Bang, but still the Big Bang might have possessed all the physicality from which our Universe is derived. And singularity might rule at the end, possibly packing all our physicality into the ultimately compressed power package. Such a state of affairs is a far cry from assuming that something came from nothing or that something became nothing. Everything that exists might ultimately undergo transformation and thus be 'mortal' – whilst nature as such may be immortal! The principle that nothing becomes nothing might be the only everlasting truth! Interestingly, such a perspective is not so far away from the ideas about Schrödinger's wavefunction being the ultimate truth, giving physicality to everything, or about the

P. Hulsroj, *What If We Don't Die?*, Springer Praxis Books,
DOI 10.1007/978-3-319-19093-8_18

mathematical universe and how only relativity as described by mathematical formulas are real, as propagated by Max Tegmark.

In this conceptual framework, the fundamental question arises whether nature, the wave-function and the mathematical universe, as alleged eternal truths, are true for the material world only, and whether, therefore, spirit, immateriality, goes by other rules. Is also spirit a part of nature, and therefore putatively everlasting; is perhaps only spirit everlasting,[1] and, ultimately, is the wave-function or Tegmark's mathematical relativity just one manifestation of spirit? If all is spirit the leap to a non-personalised god is not big (see also Chap. 13). The leap to the Great Spirit of Native Americans is not big!

Be this as it may, for the purposes of this book it is not decisive whether immortality in its crystalised, true form really exists. A billion years of life for a human is a good enough proxy for immortality, and almost all the issues addressed in the book are as relevant for the almost endless life as they are for the truly endless one.

[1] These questions echo the dualist/monist debate.

Chapter 19
The Limits of You!

In this book there is a lot of discussion of why an eternal life might not be all it is cracked up to be. Exactly the finiteness of life might be what animates it. And when faced with the prospect of becoming part of a larger fabric of consciousness, be it networked collective consciousness, machine-facilitated collective consciousness, pantheism, Buddhist Selbstaufgabe, Über-ichs, returning to God or super-consciousness, we may meet the same question of whether it is not the finiteness of our personality that makes it worth having one.

Laudably many people strive for self-improvement, and it may be argued that it is exactly this striving that gives meaning to life, at least for a large part. Our aspiration is perfection, yet our achievement will never be that, which is probably good! Perfection is something lifeless, and because lifeless something cold, as discussed above in the juxtaposition of Gilda and Rita Hayworth. Perfection might not be what we want, because it runs counter to the human condition. In fact, perfection might not be perfect, as contradictory as it may sound. In earlier times very beautiful women might have put an artificial birth mark on their cheeks to break perfection, and very handsome men might have strived for a small facial scar for the same reason. In the cult of geniuses much is made of the personality flaws that give birth to great accomplishment. Aficionados of bespoke tailoring praise the combination of perfection of fit and imperfection of the handmade. 'All your perfect imperfections'[1] might sum it up.

In our private lives we despair at our failings, but we also understand that without these failings we would not be the ones we know us to be. The cynic might wish for more sentiment, but might not wish to eradicate cynicism completely; the over-sensitive might fight against the vulnerability, yet might not wish for equilibrium of sentiment. And, in any event, equilibrium might not exist, because it presupposes the existence of an ideal human being. Human beings complement each other: the mason must not be the carpenter, but together they can build the house. The ultimate

[1] All of Me, John Legend.

P. Hulsroj, *What If We Don't Die?*, Springer Praxis Books,
DOI 10.1007/978-3-319-19093-8_19

beauty of humanity is that it is a mosaic which is rich and diverse. Strengths and weaknesses interact to eventually create a whole!

Limits of time, limits of knowledge, limits of ability, limits of experience, even limits on compassion, goodness, arrogance, vile, are perhaps what animate us as individuals! Our substance as human beings may not be defined only by our limits, but without limits would we have substance? A vase gives meaning to its water, the coasts meaning to the oceans. It may well be that our limits are what give meaning to us!

As discussed earlier (Chap. 17), we may be partly defined by the Other. Yet, we may also be defined by our understanding of the Other's understanding of us. And the face-to-face encounter with the Other might make us take responsibility for the Other, as explained by Levinas. But this all does not make us the Other! We are still finite! We should want to share with the Other, but sharing necessarily implies that you are not One. Oneness might be a dimension that we cannot grasp because of our evolutionary conditioning, but at least as an earthly concept, such as in networked or machine-based collective consciousness, it should scare us witless! Limits serve a laudable purpose!

Chapter 20
The Mind/Body Problem Resolved?

In his brilliant book 'The Protestant Ethic and the Spirit of Capitalism', Max Weber explained how the protestant ethic of duty and acquisition has led to the ideology of the pursuit of material happiness and the development of capitalism. However, in an extension of Weber's thought one may further conclude that the spirit of capitalism has also brought about an utterly materialistic perspective on immortality. To some extent this explains why religion remains such a force in a highly hedonistic society such as the United States. Religion in America often offers a contractual relationship with God, according to which God provides eternal life if the human being provides good behaviour. It is true, of course, that the pursuit of immortality has always been the object of Western religion, think Pascal's Wager, but its strong survival in America, the pinnacle of materialism and home to pockets of deep unspirituality, must be assumed to be ultimately traceable to the materialism introduced by the protestant ethic.

As little thought goes into what immortality is ultimately about, there seems to be an almost Pavlovian capitalist reflex in this immortality acquisition – in so stark contrast to Eastern religions and thought, where the sort of continued life that Westerners strive for is seen as the curse of reincarnation. Westerners want perpetual striving and acquisition – Asians seek release from desire! Given the strong Christian traditions in Eastern Europe it is small wonder that communism could not survive, and although the rampant materialism of present day China gives lie to Eastern spirituality and the rejection of desire (but, in truth, Chinese culture was never entirely unidirectional in this respect), India with its many ascetic sects might give some confirmation.

A millionaire like Russia's Dmitry Itskov is creating projects aimed primarily at giving himself immortality by 2045, again apparently without true consideration of what such immortality could be used for, but very much in the Weberian mode of acquisition. His Global Future 2045 project addresses many of the technical challenges, and has started to create an early version avatar for him, but is singularly lacking in spiritual inquiry. The most fundamental flaw in this respect is the

P. Hulsroj, *What If We Don't Die?*, Springer Praxis Books,
DOI 10.1007/978-3-319-19093-8_20

assumption that the centuries old problem of the relationship between mind and body is taken care of just by creating an eternal body that can house a mind.

When technologists speak of machine based life they tend to equate intellect, consciousness and soul, although each element merits separate consideration. With a perfect synthetic brain model it might well be that the human intellect could be replicated and that even the idiosyncrasies of individuals' minds could be reconstituted. This, however, does not mean that memory, a key feature of consciousness, can be transferred. The technologist school of thought tends naturally to adopt the physicalist point of view, according to which the mental substrate is a physical substrate, and where therefore everything stored in the brain will be replicable in a perfect model. If memory is merely signals processed in a special fashion then signal generator, signal store and processing means can be recreated. However, the immensity of the brain modelling to be able not only to think like its guest, but to be able to absorb the experience of the guest, remains a challenge that may never be met even if the physicalist dogma turns out to be correct in general.

Compounding the challenge is the fact that even a perfect brain model able to absorb all the experience of an individual, and perhaps even improving access to memory, will not necessarily have cracked the problem of consciousness, only the problem of intellect.

The physicalist will have us believe that we know, or can know, all substances, and that we will be able to replicate them. This is notoriously not so. As the Danish seventeenth century psalm suggests 'they did not have the ability to put even the smallest leaf onto a nettle', the fact remains that humans, despite strides in 'synthetic' biology, are not able to recreate organic matter, are not able to create life, except by the processes prescribed by nature: sex, planting of seeds, all the time-honoured ways of procreation.[1] Craig Venter will have us believe that he has produced synthetic life, yet the reality is that he has modified life by advanced uses of genomics. Still, he needed natural life to give it life. Even in the non-organic field we are not perfect: artificial diamonds remain distinguishable from nature's work, and gravity is understood as a phenomenon, but not as a 'substance', and cannot be produced artificially. The riddle of life is not resolved by being able to replicate physicality any more than it is by trying to furnish a dead human body with all prerequisites for life and trying to give it life. All our advanced transplant techniques still do not allow us to breathe life into a living thing after it has died. Life support systems are still a necessity to retain life; genetics appear to be no guide to life creation; consciousness is not explained by genetics and it is not obvious that electric impulses are the distinguishing feature of consciousness.

A physicalist view that nothing perishes[2] and that consciousness hence must have substance character does not necessarily lead to the conclusion that, at this time, or ever, we will gain material knowledge of consciousness, let alone the ability to tame it. Perhaps spirit is constitutionally unable to materially quantify spirit. Perhaps an assumption of consciousness mastering the substance of consciousness

[1] Although we do now know how to attach a leaf to a nettle.

[2] Similarly to re-incarnation thinking.

is a circular logic, not recognising that while humans have been incredibly successful in deploying consciousness to investigate our universe and all that is in it, including manifestations of consciousness such as love and empathy, consciousness may never be able to look inwards at itself in the manner that would be needed to make consciousness tangible. This might sound like the Christian incantation 'blessed be he who believes without having seen'; might sound as if consciousness will not reveal itself fully, as little as the Christian God. Yet, perhaps the more apt comparison might be to nature's resistance to cannibalism, as demonstrated in BSE; matter not allowing itself to be used for the purpose of the same matter. Consciousness resisting to be used to commoditise consciousness!

In addition, there is the question of soul, and whether soul is different from consciousness, as maintained so strongly by Christian faith. Thus when John Paul II recognised the validity of evolution theory, he explained at the same time, to set humans apart, that it is God who bestows soul, and that soul is not the result of an evolutionary process. At least in Roman Catholic thought, machine based life would appear antithetical to God's design, as humans would never be able to replace God as the bestower of soul. Even if humans were able to transfer consciousness, life, to machines, soul would not go with it. It may sound monist, but from most religious perspectives a separation of the 'finite infinity' of the soul, in the beautiful words of Emily Dickinson, and the vessel that was built to house it is inconceivable. The *you* you must stay in its house.

A fundamental question arises then as to whether the concept of soul is exclusive to religion. Is there something more to humans, and perhaps to other living things, than just consciousness? Many irreligious people will argue that consciousness is not all there is to self and might argue that the continuous sense of self as times and faculties change is a demonstration that consciousness, albeit part of soul, is not all of soul. Hume, and some strands of Buddhism, suggests that personality is unreal and that identity does not exist. But for most this is so counter-intuitive as to be wrong. All human beings have a feeling of centrality in their lives; a centrality that binds them together, with all their abilities, faults and experiences, even as abilities, faults and experiences change over time.[3] Again, one can debate whether this feeling of self, this soul, is just a synthesis of a variety of sensory inputs from the present and the past, organised into tangibility by our refined organism in a perfectly physicalist fashion. Or, one might suggest that consciousness is a substrate not categorised by human consciousness, and that soul is yet another substrate defying human specification and categorisation.

In the final analysis one can ask whether the immortality devotees, like Dmitry Itskov, did not start at the wrong end? Perhaps the quest for immortality should have started with exploring whether 'there is more between Heaven and Earth', whether our science until now has made us transfixed by narrowly defined physicality and

[3]The philosophical debates on presentism versus temporal parts and the like is essentially a labelling exercise trying to capture this reality. But the question is more than just a question of labels or perspective. Soul is a question of whether a different quality exists in addition to just the glue that binds together past, present and future.

has closed our eyes to investigation, beyond religion, of the nature of consciousness and soul? Perhaps there is a meeting point between the assumed first principle of epistemology and religious investigation, which will bring humankind much more understanding of what many today believe to be irrelevant or superstition. Perhaps only after such studies should we start to investigate whether the vessel nature has given us to carry spirit can be improved, whether spirit and vessel can be made immortal on earth – and perhaps we would then also understand better whether this is what we should desire.

What has been discussed until now could be said to be a mind-mind problem. Are intellect, consciousness and soul distinct qualities and if so what are their interrelationships? Moving on to the proper mind-body problem the believers in the transportability of the self might find themselves in equally uncomfortable surroundings. The starting point for machine-based life theory can normally assumed to be physicalism, mind is a derivative of matter. But that is perilous, because the whole object of the exercise is to separate mind from the matter it was naturally given. When you change matter you change mind. No problem, the transportalists might say, because we will endow our mind with better matter and hence also our mind will become better. This is, of course, possible, but from an ontological perspective is it then the same mind? In a very awkward manner the proponents of machine-based life get closer and closer to the idealist monist position, according to which mind is supreme and matter if not a derivative then something of secondary importance. Facetiously, it may be argued that transportalists get caught between the chairs, or, in fact, seek to occupy several chairs at the same time. The transport element presupposes mind as something physical but at the destination it is assumed that the self is preserved even if the physical vessel has changed. Based on exactly the intellectual premise of the advocates of machine-based life one must conclude that even if experience and sensory history can be transported, the self must change when the vessel changes. Their hope might be that the change will be only slight and only for the better.

If they ever come to an experiment they might then prove their extremist physicalist theory, but would we ever know? An avatar will never be able to tell whether it is an avatar or the original self. How could it, when the ability to distinguish would run counter to its avatar nature? From the outside, of course, it would be observable whether intellect and consciousness would be comparable to the earlier state, but would external observers ever be able to measure soul, when even the possessors of soul themselves see it only so dimly? It is popular to talk about soulless machines and, in fact, our avatars might become the supreme, but unacknowledged example.

A cutting-edge philosopher like Derek Parfit might suggest that it is an empty question to ask whether continued identity is preserved in these situations, since we know all the facts and the label is irrelevant.[4] But do we in this situation know all the facts? We probably only know all the facts if we believe that there is no mind-mind

[4] In this direction Reasons and Persons, 214, but to be fair Parfit also does not accept the mind-mind distinctions made here, so his framework is conceptually coherent.

distinction to be made, and for many exactly this distinction is the critical one, even if it cannot be externally verified. Transportalists assume that the mind will move from one vessel to another. But this seems wrong. A duplicate mind might arise, but there is no reason to believe that the you of you will move to the duplicate mind, any more than believing that all the yous arising in radical multiverse theory will be *you* you!

In sum, it is hard to see that the mind-body problem could ever be resolved without resolving the mind-mind problem. But even with the mind-mind problem resolved it would not be too likely to enable mind transportation, unless not the physicalists, but the dualists, holding that mind and matter are completely separate, are right. And then the question would remain whether a new matter would hold the old mind, when the old vessel is also still around and still houses a mind identical to the one uploaded to the new vessel!

Chapter 21
So Why We Were Not Created Immortal?

It is true to say that nature is not infallible. Even with Darwin's laws the best solution is not always found. Humankind has, for itself, improved nature's work in many ways. Seeking immortality can be understood as just one further big step in that direction.

But seeking immortality can also put into focus the question of why nature itself did not lead to the creation of immortality, or something close to it. Why is organic life so much less durable than non-organic substances, why is life more fragile than being dead? Paul Simon might be giving part of the answer in 'I am a rock': 'And a rock can feel no pain. And an island never cries'. The ability to feel and to reason has a price. One can, in a far-out way, speculate on whether the sum of sensory experience is always the same regardless of which substance is analysed; the day-fly feeling the same in one day as a granite stone over the millennia of its existence; that flesh is hence so much more fragile than bone. This chimes with our popular perception of poets burning up so quickly because of the intensity of their feeling. Yet, it is surely wrong; one tree uprooted will have 'felt' less than the tree that falls as the result of the weakness of old age; when medicine saves the life of a child it does not follow that the life of this child will be more fallow thereafter; an elephant's day is not likely to have been filled with less emotion than that of the lizard. The general proof of fallacy is, indeed, that more advanced life-forms do not tend to live shorter than simpler life-forms. And still there might be a more general correlation between intensity and life-duration; one that may explain why nature chose genetic refreshment rather than genetic indestructibility as the path to its lesser version of immortality.

Everything that feels and thinks get tired! Everything that feels and thinks amasses experience, yet experience is also a weight, partly because the useful results of experience are not the only ones retained, partly because the way to the results will be also remembered. The ability to retain knowledge becomes strained. The rationale of nature is hence that it is much better if there can be ever fresh starts – with the old generation passing down useful results to new generations unburdened by the path to enlightenment. 'Parents lend children their experience and a vicarious memory; children endow their parents with a vicarious immortality' as George Santayana put it.

P. Hulsroj, *What If We Don't Die?*, Springer Praxis Books,
DOI 10.1007/978-3-319-19093-8_21

Language is the most probate tool for passing on experience, and hence, vis-à-vis nature, language removes part of the argument for immortality. Nature's investment in language can be said to be at the expense of investment in immortality. Language allows us as a species to clear the mind-computers of useless information, and retain only that which is relevant. In this sense every generation is nature's buffer; a buffer which once in a while must be cleared. This is hardly pleasant for the generational buffer, but eminently logical from nature's perspective. After all, nature also forces us to sleep every night and we are 'reborn' every morning. Every individual wants to retain his or her consciousness, the logic of nature dictates exactly the opposite! This should give us pause for thought when we ponder immortality. When we tinker with one part of the mortality equation then we also tinker with the other. When we extend the duration of consciousness we also extend the extent of tiredness. For a fresh mind the price of new sensory impressions is easy to pay, the storage room so gapingly empty. For an old mind the benefit of new sensory impressions diminishes, and the price, although the same, appears much dearer. For a 10,000 year old mind any price might become too much. Perhaps the eternal repetition of Nietzsche is not as apposite as the fear of overwhelming tiredness. Perhaps our vessels are not limited by physical factors, but by spiritual ones. Perhaps our mental vessels can contain only so much; perhaps our reservoirs of energy can pay only for so much. Perhaps we are looking for an altogether undesirable mirage! Plato might have been right: 'But our creators, considering whether they should make a longer-lived race which was worse, or a shorter-lived race which was better, came to the conclusion that every one ought to prefer a shorter span of life, which was better, to a longer one, which was worse'[1] – or less positively Schopenhauer: 'Require the immortality of the individual is wanting to perpetuate an error to infinity'.

[1] Timaeus.

Chapter 22
Ego, Todestrieb and Immortality

Sigmund Freud explained to us our multi-layered selves, the id, ego and super-ego. If continued consciousness is the desirable element in immortality this would seem to translate into an interest in carrying forward the ego through the destruction of death. Yet, if our true origin is the id, the libido, it can be postulated that the central element of existence is lust, and that as long as the ability to lust is preserved then we are preserved. On that logic we become lesser existences when age lessens our drives. More importantly, the new-born is then the truest version of our self – as Wordsworth would have it. Preservation of our experiences through memory is no longer central to our quest for immortality, but our ability to lust is! Cogito ergo sum is then replaced by 'I lust, therefore I am'. This kind of radical ontology would unite human beings with all organic life and would in this sense be pleasing. Yet, is it true that what defines us is the ability to lust, rather than to appropriate lust, as we do through memory? Memory without lust is what computers excel at, but that hardly qualifies as an expression of self, and lust without memory would seem entirely as hollow. In Hindi/Buddhist rebirth theory lust is the enemy, the fuel that keeps life and the undesirable cycle of rebirth going, and it would hence not appear foreign to Hinduism and Buddhism to consider the id as central to the definition of self. Yet, exactly the possibility to eliminate lust and thereby move the self to Nirvana shows, of course, that the id is an eliminable part of the self, not its centrality.

P. Hulsroj, *What If We Don't Die?*, Springer Praxis Books,
DOI 10.1007/978-3-319-19093-8_22

I lust, therefore I am?

If the self is conceived of as, at least, partly id and partly memory, the dilemma is avoided of speaking about the continuity implied by immortality in the context of an element, lust, with no relation to continuity. Of course, the Kierkegaardian living in the 'moment', the fulfilment of lust, seems entirely antithetical to a desire for immortality, but then Kierkegaard was certainly wrong in suggesting that we have no interest in the past, just because it cannot be made present. In a sense we have an interest in the past exactly because it cannot be made present – because it lives forever! What defines us is not just the present, or the present and the future, it is the present, the past and the future.

Accepting that also future is a feature of the self raises another Freudian question, namely that of the Todestrieb, the death drive. Why would human beings strive both for immortality, living up the id forever, and for death, defined by Freud as the desire to return to inorganic life? Why such two diametrically opposed forces in our personality structure? In fact, Freund was never truly able to reconcile the two forces, even as he grew more and more convinced of the existence of the death drive. Schopenhauer had made it easy for himself, since his philosophical system was based on pessimism, but for Freud, the inventor of the pleasure principle, this was not easy.

Genetics might give the answer, however, since there is no principle in genetics that excludes co-habitation of contrary forces, but also on the broader conceptual level it is hardly unthinkable that we are conditioned to first seek fulfilment of our libidos and then to seek our deaths, exactly because we are not designed to be immortal (and our quest for immortality is hence perhaps an unsubstantiable dream). We might seek death as a result of life, because we do get tired, because we do fill our vessels, and when the vessels get overloaded they slowly start to sink! A ship has no desire to sink (or to float for that matter), but sink it will when it can contain no more. The death drive might be the inbred, sub-textual realisation that with time we will overload, that our libidos will have been exhausted, and hence that it might be right for us to seek oblivion, and let our children be born as new vessels – unloaded and with still untamed ids! It may be hard to agree with Schopenhauer and Freud that the purpose of life is death, but it is not hard to agree to the far more hopeful statement that both life and death have purpose.

Chapter 23
Why Immortality Is Death

When Siddhartha sets out on his quest to eliminate desire it is in a pursuit of the classical Hindi/Buddhist idea that through this Nirvana would be reached. However, there are two perspectives on eliminating desire, the Hindi/Buddhist and the aspirational, and in a sense Siddhartha ends up pursuing both, one after the other, only to end with having had fulfilment from both. By first abstaining and then indulging he gets to the Zen, requiring the simultaneous acceptance of all the opposites of existence.

The traditional Hindi/Buddhist perspective is, of course, one of understanding desire as undesirable. Desire is a burden and the sooner we get rid of it the better; the sooner we escape the cycle of rebirth, which is not a present of ever new glory, the sooner we escape lives that appear to be punishment. The central message is that life is not a many-splendoured thing – life sucks!

The aspirational perspective is the inverse. Life is a true present no matter how much hardship is attached. Life's difficulties are there to be conquered, that is true, but in the final analysis life is an opportunity from which as much enjoyment as at all possible must be had. Materialism springs from this well, as do sex-obsession and ultra-aestheticism.

To the extent one can make a choice between these two alternative ways of looking at life, clearly the more useful (if usefulness is a criterion) and satisfying is the aspirational. In fact, the aspirational approach fits wealthy societies hand-in-glove. However, in the final analysis the fundamental difference between the Hindi/Buddhist and the aspirational paths is not necessarily desire as such, but how desire is handled. Wealth-inspired aspirational societies believe that desire must be satisfied to the hilt to get rid of it, poverty-inspired Far Eastern societies that desire must be overcome as there is no viable way to satisfy it. Paradoxically, for many on both sides of the divide the destination is the same, namely that desire will no longer be the prime mover. The Christian Paradise is also one free of want, perhaps less clearly sketched than Nirvana, but having a similar topology.

In many respects the Hindi/Buddhist understanding is a life version of the Hades of the ancient Greeks. Even before death, or rather through all the repeated deaths,

P. Hulsroj, *What If We Don't Die?*, Springer Praxis Books,
DOI 10.1007/978-3-319-19093-8_23

you live a life in the shadows and joy has little place. The immortality flowing from both Far-Eastern and Greek religion is thus frightening. In Hades the shadows are ever-present. In the search for Nirvana or release you walk though an endless chain of valleys of tears until at last you reach Atman and unity with the Absolute, in effect complete Selbstaufgabe. Nirvana is perhaps better than nothingness, but not so different. The raw material returns to its source. All the lives of a person might possibly be understood as a kind of purgatory bookended by Nirvana at both ends! Yet, an existence in Hades is much worse than nothingness, although Hades in contrast to Hindi/Buddhist thought might serve as an admonition to the living to enjoy while they can – so some utility attaches!

The aspirational belief contains, of course, something far more profound than just materialism and a craving for sex. Aspiration is the fuel of life. Evolution has conditioned us to strive and some of the West's most beautiful music, literature and art express a most profound sense of longing. And although longing has only as a *premise* the possibility and the attractiveness of arriving – the greatest joy is the joy of expectation[1] – it is true to say that our kind of life must involve the reality that the reward of aspiration, achievement, is truly desirable, not just imagined as being so.[2] Conversely, with nothing to strive for there can be no life in any earthly or temporal form. Rewards without desire for the rewards will not do the trick. What this means is that when desire no longer burns then we stop living, as Hinduism/Buddhism teaches. We then reach Nirvana, a place of 'rest'. The conundrum for earthly immortality is that if you have immortality to what would you aspire? Over a longer or shorter period you will have achieved all within your possible reach and the fuel of life, desire, will have run out. The concept of immortality is possessed of an internal contradiction. With immortality you will ultimately want to die, and you will most likely die as a result, unless atemporal eternal life has eliminated this deep-seated personal need for Tao. But unfortunately atemporal eternal life is not achievable for humans this side of the Styx!

If one assumes that the principle of communicating vessels is a universal governance principle, that matter is always matched by anti-matter, joy with sorrow, that a cosmic balance exists in which every transformation is counterweighed in such a fashion that equilibrium is retained, yin and yang, then moving the lever of immortality to the right will mean that the lever of death will move to the left. Humans would only gain immortality by introducing more death elsewhere or, perhaps, the balance will be maintained by immortals voluntarily seeking death! At a very basic level, humans are animated by the struggle against death, and if that struggle is swept aside then humans will no longer be animated.

The Germans, and possibly Confucius, say 'Der Weg ist das Ziel'. If the road becomes endless it can no longer serve as a destination, and existence might become pointless.

---

[1] Attributed to Soeren Kierkegaard, but origin uncertain.

[2] In stark contrast, Giacomo Leopardi, Saturday in the Village, who seems to suggest that the objects of desire will mostly, perhaps always, be disappointing, and that therefore the uplifting spirit of expectation is entirely hollow.

Chapter 24
Why Seize the Day?

With an endless string of days ahead of you, why would you seek to make something special out of any specific day? After all, the billionaire appreciates each individual dollar less than the pauper. This does not go to say that the billionaire treasures money less than others, but the billions have made him free to think about the use of money in a broader sense. One might suggest that the same could become true for those who would possess immortality: the individual day might become less important but the wealth of days would allow approaching time with a longer term investment perspective. The scent of the evening rain might become boring but, for that, the satisfaction of having perfected the tennis backhand over a thousand years might be so deep.

Sadly, our hard-coded human nature might disallow the satisfaction of the long-term investment perspective, whilst dictating diminished return on the scent of the evening rain. It would seem that our nature is bound to a temporality that is relatively instant. It is true that we invest much life-time in education in order to reap long-term returns, but even there we tend to look to instant gratification as well. That so many well-educated people are so nostalgic about the time of their studies might be strong evidence that the time of education was not only one of sacrifice. Even the ascetic monks, the wearers of hair-shirts, the saints on columns might not be investing their lives just for the long-term. Pain has its own aesthetic, its own appeal to the instant.

US corporations are always criticised for investing for the short term. Humans are possibly not so different from US corporations. The exhortation of Horace to seize the day might be less of an invitation to a paradigm shift in how we feel and act, and more an invitation to become even more urgent in how we feel.

When John Kerry was the Democratic presidential candidate in 2004 it was noted that he once answered a question on what he would change by himself if he had a free choice with 'remove aloof'. But even Horace would probably accept that those who are aloof, those who always have their heads full of plans for the tomorrows, do this in order to also satisfy today's feelings. Just like nostalgia can be an incredibly strong feeling, so can the planning for tomorrow. The catch in this is that

P. Hulsroj, *What If We Don't Die?*, Springer Praxis Books,
DOI 10.1007/978-3-319-19093-8_24

even planning for tomorrow will start to lose its appeal if the future holds a never-ending number of tomorrows. Why seize the day – why seize any day?

In God's realm immortals may be possessed of a sensory apparatus allowing unlimited enjoyment without reference to time, atemporal enjoyment even, but on Earth it is hard to see how the link between joy and time could ever be broken.

The relativity of feeling to time is beautifully, if harrowingly, illustrated by a poem by Tom Kristensen:

The Execution[1]:
....
We kneel, we twenty men
We have our heads outstretched
And I must see bright steel
Strike off the heads of five
But now the sixth, the sixth…
My time goes deadly slow
The eye is blank
And everything is done

The headsman now begins
To polish fourth time round
On that same rag the sword
While number four slumps down
And blood flows in a gush.
The headsman steps up close
I focus on the hilt
There is a dragon's flourish
On the crossguard of the hilt.
....
Did the world come to a stop?
The sword – is it still moist?
Will the headsman always polish it
And never put it to use?
My nape stings constantly
The pain encircles it
A ring around my neck.
Is it that I am dead?

No, the man stands still and looks
Along the sword's hard edge
Then makes another step
And stops, and calculates, steps back.
A beetle trundles safe
Green metal on curved back;
It walks serenely towards
An executioner's foot.
....

Horace would understand, but do we?

[1] Translation Harry Eyres.

Chapter 25
Loss and Value

As shown, the magic of the scent of the evening rain might be appreciated less and less with time, and might be entirely lost with eternity.

This has led some philosophers, such as Samuel Scheffler, to conclude that the human value system would be upended by immortality. Little of what we now cherish will have value once we know that we will have endless opportunity to take benefit. Professor Scheffler gets surprising support from the pop singer Passenger:

> Well you only need the light when it's burning low
> Only miss thc sun when it starts to snow
> Only know you love her when you let her go
> Only know you've been high when you're feeling low
> Only hate the road when you're missing home
> Only know you love her when you let her go

Allegedly, it is loss and death that inform our appreciation and our values!

This logic is not quite as straightforward as all that but is akin to the one suggesting that great art or achievement can only come about through the crucible of suffering. A logic which is particularly repulsive when it makes idolatry of suffering and death, as so much Russian literature and culture, for all the intensity and beauty, have tended to do.

But surely there is a more nuanced way of understanding the background to our values, at least as regards the effect of immortality on our aesthetic. If we know we possess eternity we also know that first time encounters with experiences and impressions are a limited commodity. It would seem likely that this realization would mean that we would become even more consciously appreciative of any truly new stimulus. Imagine the delight if after a billion years you see a whale jump for the first time. Like a glass of champagne in the desert!

Of course, this does not take us away from the tedium of repetition and from the Makropulosian coldness that comes from having seen almost everything so many times before. And our senses will numb with excessive time so that, perhaps, even the first face-to-face with the whale will be lost on us if it only takes place after a billion years. Yet the point is that numbing takes time and if we know that we have

P. Hulsroj, *What If We Don't Die?*, Springer Praxis Books,
DOI 10.1007/978-3-319-19093-8_25

eternity we will also be keenly aware of the numbing – and the need to delight as long as we can. The numbing becomes the loss that would inform our values and the related awareness of the ultimate numbness and loss of feeling would be the driver for being appreciative for as long as possible. If you know your mind will petrify perhaps you will make efforts to squeeze as much as possible out of your ability to feel whilst you have the ability, and you may make the utmost effort to delay the onset of the numbing. In this sense there may be little difference between the immortal and the Alzheimer patient!

In the final analysis this might show that physical and spiritual immortality are unlikely to go hand-in-hand, and that what we should fear is not physical but mental death! It is the prospect of gradual mental death that should then inform our values!

Chapter 26
Are the Gods Immortal?

Viking gods are fantastic in many ways. One of these is how they represent a supercharged version of humanity. Viking gods fight, drink, love and die, and in the end evil is stronger than good and the earth, Asgard, all but two of humankind, and most gods disappear in Ragnarok, a Viking version of Noah and the flood. Immortality is not an inevitable part of the Viking deal, being a god might bring you a wilder and longer life but not an eternal one, even if some gods, and the two humans, move on to a rebirth of Earth and godly domains.

In other theistic religions divine immortality tends to be the standard. But still, in Greek, Roman and Hindi religions being a god is not an absolute guarantee of eternal life – just consider in Hinduism the beheading of the god Ganesha and the transplant of the elephant head to reawaken him to life. Where these religions are most similar to that of the Norse is in the imperfection of the gods. Greek, Roman and Hindi gods share imperfection with humanity. The Judeo-Christian religions are different because god becomes impersonal. God's message, his ideology, is highly personal, but god himself is perfect, does not share humanity with us, except, in Christianity, through the son of god, and even there the shared humanity is largely the shared earthly mortality.

There is logic to the immortality of atemporal gods, and no logic to generational succession of atemporal divinity. There is possible logic to generational change-over for sempiternal divinity, but, nevertheless, lots of conflict potential in inter-generationality if god is temporal but immortal. Will a son be caught in the role of son for eternity in that case? In Christianity the trinity doctrine might be assumed to resolve this apparent conflict, but even so why use the concepts of father and son if no succession will ever take place? 'Sitting by the right hand of God Father the Almighty' is complex symbolism when perfection, eternity and monotheism are sought to be projected.

It might thus be tempting to speculate on divine generational renewal in more general terms, particularly if eternity is understood as sempiternity, and ponder why the forces creating the Creator would have followed a different generational logic than that enshrined in evolution. Albeit interesting, such a line of argument would

P. Hulsroj, *What If We Don't Die?*, Springer Praxis Books,
DOI 10.1007/978-3-319-19093-8_26

A god dies!
(Death of Baldur)

probably miss the point, however. Because the point might be that eternal life, be it atemporal or sempiternal, presupposes perfection. Viking gods were imperfect and hence still ruled by death – Greek, Roman and Hindi gods certainly imperfect as well, and not completely insulated from perdition. It is only with the Judeo-Christian religions that god becomes impersonal and therefore perfect and therefore absolutely immortal. The lesson for humanity's quest for immortality might be that unless we believe that endless time will bring us human perfection then earthly immortality might always be beyond our reach. Our imperfections, as likely to be exacerbated as cured with time, will ultimately always take us to Ragnarok.

Chapter 27
Immortality of the Devil

In our feel-good times it is easier to believe in god than in the devil. Yet, Loki was stronger than Baldur, and evil led to Ragnarok. Given also our putative genetically induced egotism perhaps it should be easier to believe in the devil than in god, or at least to believe that if god exists then also the devil must exist. As Karen Blixen explained: 'God and the Devil are one, the majesty coeternal, not two uncreated but one uncreated, and the Natives neither confounded the persons nor divided the substance.'[1]

Again, if we believe in eternal balancing and thus that something must always be counter-balanced by something else, a belief in god must lead to a belief in the devil, a belief in the immortality of god to a belief in the immortality of the devil, and a belief in the immortality of good to a belief in the immortality of evil. The wishy-washiness of moral nano-differentiation invites ultimate moral indiscrimination and rejection of the stark choices of good and evil, yet even if human action is, in fact, often morally grey, and should be understood as such, there is no invitation to forget that grey is a composite of white and black. In a perverse way there is an eternal morality of evil, which must be eternally fought, even if that fight involves love as the weapon of good – love of fellow humans, however, not love of the enemy, evil.

Immortality of the devil means, of course, that earthly eternal life will also always be accompanied not only by god, but also by the devil. And in an irreligious sense, the immortality of evil means that earthly immortality will always have to confront evil, not only rejoice in good. That is the fundamental difference between divine immortality and the earthly variety: if you reside with god, the devil has been vanquished, if you reside on earth the devil is ever alive and kicking!

[1] Out of Africa, 28.

P. Hulsroj, *What If We Don't Die?*, Springer Praxis Books,
DOI 10.1007/978-3-319-19093-8_27

Immortal, moi?
(Henry Fuseli, The Nightmare)

Recently it became quite fashionable to discuss whether human beings possess a God gene. This is, in principle, an interesting topic, but perhaps we should not only reflect on genetic explanations of religion, but also on genetic explanations of the human being's ability to do good and to do evil. If altruism is a mirage (and this can well be doubted) and altruistic behaviour only motivated by genetic propagation considerations, and hence evil motivated by hard-coded egotism, who knows whether our dissection of genetic codes will eventually lead us to be able to reinforce our ability to do good and decrease our readiness to do evil. Perhaps the devil resides in our genes and perhaps he can be smoked out. Perhaps there is a gene of good and a gene of evil, and perhaps genetics will give us an even greater gift than immortality, namely the gift to exorcise our readiness to do evil. No greater social engineering feat is imaginable!

Obviously, this kind of social engineering is fraught with danger, inter alia because there is, paradoxically, an element of egotism even in love and altruism. And this is true even if you do not subscribe to the crasser versions of evolution theory. Uncomfortably, good and evil are more interrelated than we would like to think, and eliminating elements that are prima facie evil might unintentionally eliminate necessary elements for doing good. The devil might not be god's neighbour, but god's most problematic and most lasting tenant! Or, perhaps, big perhaps, with Nietzsche: Everything good is the transmutation of something evil; every god has a devil for a father.[2]

[2] Friedrich Nietzsche, Sämtliche Werke: Kritische Studienausgabe, vol. 10, 195.

Chapter 28
The Immortality of Not Being Born

If you assume that our souls are everlasting, both before we are born and after we die, or if you believe that spirit is a substance like gravity or dark matter (but less explored by the natural sciences), indestructible but mouldable into different forms, the consequence might be that possible life as a human being is largely irrelevant given the enormity of eternity and the flash constituted by human life. It might be said that you or what you are made of beyond skin and bones have defied death without ever having been born. In a realm of rebirth it might be unimportant if one misses the train of birth once in a while. But then, of course, if spirit is a substance or if we are part of an eternal string of rebirths until Nirvana is reached, it is not possible to miss the train, the substance is always deployed, rebirth will always follow rebirth, even if this might be in different worlds or dimensions than those we know. Immortality means always having taken some form or other. In one form dayfly, in the next twenty thousand days as a giant tortoise, and then, who knows, for one beautiful summer, a rose!

The point here is that this kind of immortality, or in reality endless mortality, would appear to make each individual life, each individual new form or rebirth, of little consequence. But this is not how we are conditioned to think, neither human nor animal. Life appears to us immensely precious as, indeed, it is. Perhaps this again points us to the relevance of consciousness, or each individual instance of consciousness, for the understanding of self. But it might also be that our attachment to our self shows that only one self exists and that our self is created of nothing only to revert to nothing. Perhaps it shows that nothingness is the eternal force. That a stone does not moan when it is ground to destruction shows perhaps that the elements of the stone are immortal, dead skin or amputated limbs as well. Perhaps only that which feels is ephemeral; perhaps feeling lives and dies, but nothing else does? Perhaps Ibsen was completely wrong: perhaps only that which can be lost will be lost! Which again can be said to mean that beyond skin and bones only loss is eternal; that the unique feature of living organisms is the ability to lose! As we still remember Shakespeare paradoxically he would not yet reached have reached eternity then.

P. Hulsroj, *What If We Don't Die?*, Springer Praxis Books,
DOI 10.1007/978-3-319-19093-8_28

Nothingness as the defining feature of living things chimes well with our times of extreme value relativity as it puts something absolute as the end point of the relativity. All relativity requires anchors of absolutes in order to distinguish itself from complete rule abolition and anarchy. Perhaps the only real anchor of our questioning of all values is the ultimate release into nothing!

In its own round-about fashion, finality has an element of eternity. Nihilism has its own Nirvana.

Chapter 29
Oh, Jerusalem!

Jerusalem is probably the single most popular place to be buried. The reason for this is neither beauty, nor history, but the allusion in all three Abrahamic religions to Resurrection, and how resurrection will start in Jerusalem. Being buried in Jerusalem, at best as close as possible to the Temple Mount or the Holy Sepulchre, will mean resurrection right when God's Kingdom Comes. Interestingly, this very literal interpretation of Resurrection implies that the dead will 'sleep' until Kingdom Come. Thousands of years of sleep might not be as frightening as sleep with no end, no matter what Socrates tried to convince us, but still there is some tension in the idea that a human goes through the death experience, sleeps, and reawakens. This is not too far from the general rebirth thinking of Hindus and Buddhists, although supposedly in resurrection your self will be unchanged. It is a conceptually tall order from many perspectives, but in terms of immortality it breaks its own logic in at least one sense. The immortal soul is put in abeyance until the last day, which means that the soul is created upon birth, but the immortal soul only upon resurrection, since only after Judgment it is known whether your soul will be chosen. And awaiting that decision you are, in fact, dead. The longing for the Apocalypse is explained by the wish to curtail the waiting time, but this, of course, opens the question whether eventual resurrection is a comfort no matter the waiting time. Being dead for a billion years and then being brought back to life, sounds much less appealing than dying and being resurrected a few years later. There are good reasons why most religious thought no longer assumes resurrection only upon the Apocalypse. Nevertheless, resurrection puts into focus that there is much relativity in immortality, and, more importantly, it questions very well our motivations for seeking immortality: is it really a wish to continue living, or is it only a fear of perishing? These are two entirely different things, and it is perhaps interesting to note that so little thought has gone into what eternal or unending life would really mean, whereas there is no paucity of thought on what it means to perish.

P. Hulsroj, *What If We Don't Die?*, Springer Praxis Books,
DOI 10.1007/978-3-319-19093-8_29

Waiting for the last day…
(Credit: David Bank/AWL Images/Getty Images)

Jerusalem has this fascinating quality of being both a physical reality and a spiritual emblem; Jerusalem is also a heavenly city. Which illustrates, again, how much we transpose earthly experience to the realm of God. When John Paul II died a cardinal told the press that the deceased Pope now had his first conversation with God. In other words, life on Earth is moved to the heavens. Such an idea of immortality appears highly unconvincing, and from a theological perspective it is surely sounder to argue that life in God cannot be divined than to assume that everything just continues, but better and with a more immediate presence of God.

Chapter 30
The Twilight Zone Between Death and Immortality

The famous Liverpool football coach Bill Shankly once said: "Some people believe football is a matter of life and death, I am very disappointed with that attitude. I can assure you it is much, much more important than that".

In a similar vein, De Gaulle pointed out that 'The cemeteries of the world are full of indispensable men', illustrating both what high priority we put on work, and how wrong life can prove us.

One of my old teachers once told me that he continued smoking although his much loved wife was very sick from lung cancer. His explanation was not lack of will to stop, but simply that smoking was a part of his quality of life and he was not ready to give it up.

Smoking as life quality at the expense of life time, like excessive drinking and drug use, is the simplest mortality equation: push back the thoughts of death by bringing it closer – no obvious quest for transcendence there. But work and football go beyond this paradigm. Sports and work ground us in the present, that is true, but sports and work often also represent a search for transcendence, like love does, ex hoc momento pendet aeternitas.

Sports and work might not bring us the communion with the eternal as sensual love might on those rare occasions of glimpsing The Promised Land, yet they still represent our wish to go beyond ourselves. The intensity of our feelings is channelled into activities that we prioritise so highly that we feel that we move beyond our shackles. A football hooligan would hardly subscribe to this theory, but even for him the primitive tribal urges that are at play translate into the tribe becoming more important than the self. Hells Angels sometimes die for Hells Angels. Love of country makes the possibility of eternal perdition bearable.

It can be argued that this reaching beyond is no different from what we seek in books, plays and movies, in fact, what we experience every time we do not gaze at our own navels. Any activity that occupies the mind with something other than you has transcendental quality.

The film When Harry met Sally has a scene in which Harry explains how he thinks about death for hours and days so that he is prepared for it when it comes,

P. Hulsroj, *What If We Don't Die?*, Springer Praxis Books,
DOI 10.1007/978-3-319-19093-8_30

apparently disregarding Shakespeare's dictum: 'Cowards die many times before their deaths. The valiant never taste of death but once'. Sally's response, in the same vein, is the pithy 'in the meantime you are going to ruin your whole life waiting for it'. Navel gazing and immersion in the immediate nicely juxta-positioned, with Sally positing that she is a happy person. Paradoxically then, the immersion in the immediate becomes transcendence; shopping becomes a tool to forget yourself, or to attach yourself to something beyond yourself. Shopping might be a Kierkegaardian 'moment', notwithstanding how much Kierkegaard would hate such a prosaic perspective.

But also love of history is a most poignant way to transcend, a way of relativising the own existence and attaching the self to events that, by definition, cannot be turned into current personal reality. Looking backwards is, in some senses, a way to not look forward, even if many a look back aims at learning history's lessons as a guide for future action. Looking back in time to defy time is full of tension, but still a way for many to transcend and overcome the personal exposure to both the present and the future. Backwards transcendence is full of beauty. Dealing with history can be said to give us god's perspective. Sitting in god's chair is a comfort that makes us partake in god's immortality even if only in passivity and, ironically, for a while!

Chapter 31
Movement and Immortality

Disingenuously perhaps, it might be suggested that there are two archetypes for how humans organise their lives to be in touch as much as possible with the aspiration for eternity. The first is to stay as put as possible throughout life, thus gaining a feeling of permanence, which might show the way to immortality. The second is to be on the move as much as possible, and thus through permanent movement gain a feeling of ease of displacement, physically as well as mentally, which might be felt to facilitate also the move from earth to paradise. Transient existence as a path to the great transition!

Immobility as a tool to ensure permanence is psychologically not hard to understand: you put down roots as firmly as possible, and hope that those roots somehow will be strong enough to overcome even the pull of death. The comfort of knowing intimately your surroundings leads to the hope that this comfort in some way or another will follow you even to the beyond.

The logic of the restless is the inverse. By becoming masters of change the hope is that also the ultimate change can be mastered. The restlessness might in the first instance be predicated on the wish to escape thoughts of perdition, might be a crude version of ex hoc momento pendet aeternitas, with the immersion in the immediate being a way of touching immortality. But underneath and in parallel to that, the ability to face continuous change might give confidence that whatever life's end might bring it will be somehow manageable.

For the restless, Einstein's relativity theory might find new meaning. If you move fast enough relative to death, death will not be as big.

P. Hulsroj, *What If We Don't Die?*, Springer Praxis Books,
DOI 10.1007/978-3-319-19093-8_31

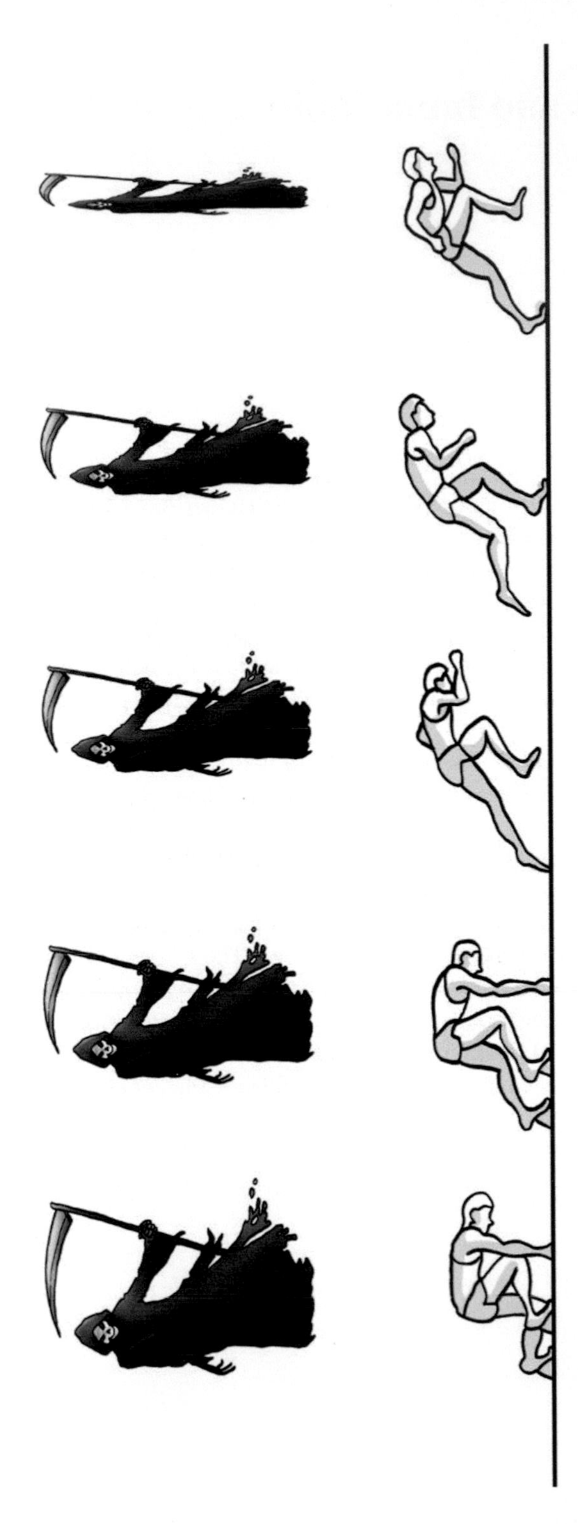

The relativity of death
(Credit: Joris Thys)

Chapter 32
So, What Is So Horrible About Death?

Our earth existed for 4 billion years before any human was born, and will exist for billions of years after humans of our age will have died, bar earthly immortality. So we can ask, as also done earlier in this book, why is it that we are unconcerned about the billions of years the earth spent without us, but are so terribly worried about our absence in the billions of years that will follow our life-times? One reason is, perhaps, that the past was never within our reach, but the taste of the future is on our lips, as if our current lives are but an appetizer for a much bigger meal. Or, more prosaically, clearly humans find it difficult to let go of a good thing. Christian religious lives in affluent societies are often inspired by the fear of death, by the fear of the loss of the privilege of life, whereas Christian religious lives in subsistence societies might see Paradise as the destination and death as the key that will let you enter and give you release from a life of never-ending hardship. The paths might thus be different, and the wish for eternity as well. The affluent might just wish for continuation of affluent life, the impoverished for a discontinuation of anguish. Still the wish for continuation of the self is shared.

Animal lives are not similarly informed. Certainly, animals know how to fear death and, to varying degrees, to calibrate risk. But active avoidance of death is the order of the day, not moody deliberation of death sometime in a more distant future. The immediacy of animal life is a key discriminator from human life. Kierkegaard reminds us that the lilies of the field and the birds of the air are always present in the moment of their lives.[1]

Peter Singer has in his ugly thesis on the un-sanctification of human life argued that because a new-born baby lives in a higher degree of immediacy than, say, an adult chimpanzee, the life of a new-born human should not be protected to the degree of that of an adult chimpanzee. In setting the ethics of evolution aside, protectable life according to Singer becomes a question of ability to have advanced feelings and forward planning capabilities, the joy of expectation you might say. This element of forward planning and sense of expectation is also the element that

[1] More of Kierkegaard on this in Chap. 35.

P. Hulsroj, *What If We Don't Die?*, Springer Praxis Books,
DOI 10.1007/978-3-319-19093-8_32

imbues death with so much horror in humans; so much horror that many forget to treasure the life we know we have been given – albeit without guarantee of duration and, even less, of immortality.

It is important to note, however, that the unpleasantness of death is different from an aversion to see our temporal existence extinguished. Humans are conditioned by evolution to strongly resist the death experience. This is true even for the slave and the disenfranchised for whom death is also understood as the key to a better life. And it is true virtually regardless of how strong the religious faith is. Shakespeare has it down best, of course, in the soliloquy of Hamlet, of which we tend to know only the first line:

> To be, or not to be, that is the question:
> Whether 'tis Nobler in the mind to suffer
> The Slings and Arrows of outrageous Fortune,
> Or to take Arms against a Sea of troubles,
> And by opposing end them: to die, to sleep
> No more; and by a sleep, to say we end
> The Heart-ache, and the thousand Natural shocks
> That Flesh is heir to? 'Tis a consummation
> Devoutly to be wished. To die, to sleep,
> To sleep, perchance to Dream; Aye, there's the rub,
> For in that sleep of death, what dreams may come,
> When we have shuffled off this mortal coil,
> Must give us pause. There's the respect
> That makes Calamity of so long life:
> For who would bear the Whips and Scorns of time,
> The Oppressor's wrong, the proud man's Contumely,
> The pangs of despised Love, the Law's delay,
> The insolence of Office, and the Spurns
> That patient merit of the unworthy takes,
> When he himself might his Quietus make
> With a bare Bodkin? Who would Fardels bear,
> To grunt and sweat under a weary life,
> But that the dread of something after death,
> The undiscovered Country, from whose bourn
> No Traveller returns, Puzzles the will,
> And makes us rather bear those ills we have,
> Than fly to others that we know not of.
> Thus Conscience does make Cowards of us all,
> And thus the Native hue of Resolution
> Is sicklied o'er, with the pale cast of Thought,
> And enterprises of great pitch and moment,
> With this regard their Currents turn awry,
> And lose the name of Action. ……[2]

[2] Hamlet, Act 3 scene 1.

It is, however, not so obvious that this fear of death and the unknown is also a fear of losing out on a future of billions of years of further existence. Life brings great joy to many people, some joy to even more, seeming indifference to some, and, as Shakespeare stresses, great pain to quite a few.[3] For those who get up every morning full of pain and fear it is hard to believe that a sempiternity of this sort of life would be desired and those leading indifferent lives could perhaps be assumed to be indifferent to never-ending indifference, but would more likely be horrified of ever-lasting grey, where even those events providing rays of light – the birth of a child, the coming of spring – have been either eliminated or become routine. And even the joyful may, on reflection, find that what provides joy is steeped in temporality, and that Nietzschean repetition might be undesirable. The praise by Socrates of the virtues of being dead – the dreamless sleep versus a great king's most pleasurable day – might not have gained a lot of traction, not because it was wrong per se, but because Socrates, like Epicurus a hundred years later (Non fui, fui, non sum, non curo),[4] did not distinguish between the death experience and the consequence of being dead. The image of sleep without dreams might not be so abhorrent, even if humans will never understand their own non-being. But what colours all our thinking of non-being is the step leading to non-being; the death experience.[5] That step can never be made appealing, no matter how exploited, pained or exhausted a human being may be.

> Grave men, near death, who see with blinding sight,
> Blind eyes could blaze like meteors and be gay,
> Rage, rage against the dying of the light.
>
> And you, my father, there on the sad height,
> Curse, bless, me now with your fierce tears, I pray.
> Do not go gentle into that good night.
> Rage, rage against the dying of the light.[6]

---

[3] Many appear, in fact, to be caught in the twilight zone described in the Robbie Williams song Feel: I don't wanna die, But I ain't keen on living either.

[4] I was not; I was; I am not; I do not care.

[5] Samuel Scheffler is on the track towards this distinction in Death & the Afterlife, but does not quite get there having become transfixed by the Epicurean logic of 'So death, the most terrifying of ills, is nothing to us, since so long as we exist death is not with us; but when death comes, then we do not exist. It does not then concern either the living or the dead, since for the former it is not, and the latter are no more.' Epistula ad Menoecerum in The Extant Remains. Nick Cave in Immortality is only the most recent to join that band wagon. Note, however, that Epicurus starts out by talking of 'the most terrifying of ills', so the distinction was not completely lost on him. Niko Kolodny's commentary on Scheffler, pp 163–4, equally recognizes the 'passage' element.

[6] Do Not Go Gentle Into that Good Night, Dylan Thomas.

Non fui, fui, non sum, non curo

Humans are reconciled with not having lived the pre-birth past, may be reconciled with not living an unending future, but will never be able to move from being to non-being without horror, unless it happens surreptitiously! Wernher von Braun is reputed to have expressed satisfaction with dying slowly from cancer, since it gave him an opportunity to observe the process of dying first hand, but not many would share this extremely scientific approach to their own demise. Routinely people respond that dying in their sleep is their preferred death, which in the final analysis is death without the death experience. Sleep followed by eternal sleep is not so far from the idea of Socrates, and, crucially for many, eliminates the horrors of consciously moving from one type of sleep to the other.

For others the thought of passing from temporary to eternal sleep without this being marked or noticed is in itself a horror, simply because many humans like the illusion of continuity. Despite all human experience of sudden death, many people believe that death will not come to them as a thief at night, but will give notice. And thus they believe that as long as notice has not been given all is well, and can be expected to remain well. Tomorrow will follow today!

Whether an individual hopes for notice or not is in the final analysis largely a personality question, and humans can influence their destiny in this respect only little. What is less of a personality question is that most of us prefer to die after a fulfilled life. That is true even for the thrill-seekers courting danger, since they tend to seek confirmation of life by challenging death. Notwithstanding Dylan Thomas, the beautiful words of the Danish poet Frank Jaeger thus ring true to most of us:

> If only
> One could be an apple
> Grow richly round and at ease
> Suck deep from the sap in the stem
> Then let go on a late summer's day[7]

---

[7] Vaeredigtet (The Poem on Being), Samlede digte, 66 (translation Harry Eyres).

Also Shakespeare, King Lear:

> Men must endure
> Their going hence, even as their coming hither.
> Ripeness is all.

Chapter 33
Lust for Life or Fear of Non-existence?

An important distinction is missed if the emotional spectrum of death is seen only as containing the horror of the death experience versus missing out on further existence. In that pairing it might well be that the horror of the death experience wins out, as argued above. But what about the pairing of the lack of lust for life versus the fear of non-existence? Or rather, the combined front of the horror of the death experience and the fear of non-existence opposing the lack of lust for life?

It may be easy to confuse a lack of lust for life with an indifference towards non-existence, but this is wrong. Psychologically, the two emotions are entirely different. Humans might have no real interest in extending the human condition endlessly, might not have truly conceptualized what everlasting life implies, so might, in this respect, not be so unsettled. After all, how can you be truly unsettled by not obtaining something you have not truly understood or wished for?

Losing something you have is something else altogether.[1] Humans are certainly not evolutionarily conditioned to understand or to welcome non-existence, even if they do not crave eternal existence. Humans do not want to lose the individual moment even if they do not aspire to an endless string of moments. Kierkegaard seeks to make us focused on the importance of the individual moment, but Peter Singer tells us that what makes a human life protectable is the ability to live out of the moment, to have anticipatory ability. An interesting dichotomy in all of this is, however, that exactly the 'living in the moment' makes the thought of losing future moments unbearable.

It could then be assumed, perhaps, that the more remote a human is from the pulse of his life the more the human should be able to accept the thought of non-existence. But even this is not true, as the Singer logic shows. Those who are more anticipatory, those who live less in the moment, live lives that will often be permeated by the joy of expectation, so obviously also this perspective leads to the thought of non-existence being abominable, simply because the time-horizon of

[1] Victor Hugo: It is nothing to die. It is frightful not to live.

P. Hulsroj, *What If We Don't Die?*, Springer Praxis Books,
DOI 10.1007/978-3-319-19093-8_33

anticipation is rather limited. No forward-looking human has a 10,000 year perspective, and exactly for that reason non-existence becomes an enemy.

In sum, whether a human being lives in the moment or lives a life of anticipation, the thought of non-existence will always be highly unappetizing.

Chapter 34
Immortality and Intergenerational Justice

The concept of intergenerational justice has a lot of wind at its back – at least until action has to be taken. There is a compelling logic to suggesting that we should not leave the Earth and our societies to the next generation in a worse state than when we took over. And there is even a Darwinian logic involved: what does it help to pass on your genes, but leave the holder of your genes to circumstances that will make the further passing on of the genes very difficult. Thus intergenerational justice can be understood also in a narrowly egotistical fashion, as a concept that protects our only current earthly immortality, our off-spring.

With the achievement of our own proper earthly immortality the question of intergenerational justice changes character – might, in fact, completely disappear as we elbow out all future generations by staying at the table forever.

It can, of course, be discussed whether we are not already in a rapid process of limiting access to Earth for the unborn. Some may argue that the opposite is the case, that with seven billion people we have forced Earth to become much more hospitable to many more human beings than ever before. There is truth and lie to this. Yes, Earth is host to many more than before, but is also host for a longer time, so the turn-over is much slower. With average global life times doubling over the last 100 years the turn-over rate is halved, so seven billion equates to a global population of 3.5 billion 100 years ago. As we become ever older we become ever less generous to new generations. In Germany and Japan birth rates are below replacement rates, and the one child policy of China has been a successful attempt at keeping birth rates in check whilst life expectancy was going up. The Malthusian conclusion is that with ever-greater life expectancy there will be less and less room for the next generation and with immortality there will be none.

Corking the generational bottle is not an issue if you believe in the absoluteness of the present. Justice is then only a concept for the living; of no relevance vis-à-vis the dead or the unborn. Historical justice is an idle pursuit and the unborn an abstraction only. A belief in the absoluteness of the present is a belief in the absoluteness of strings of moments, each independent of each other. Such a belief is hence a transposition of Hume's personality theory to society as such. When justice has to

P. Hulsroj, *What If We Don't Die?*, Springer Praxis Books,
DOI 10.1007/978-3-319-19093-8_34

be dispensed only for the moment it loses its predictive power and relevance for the future. Everything becomes ad hoc and normativity becomes impossible. What might be just with reference to the past becomes unjust when measured merely by the present. You may live a life of great comfort because of your achievements of the past, but judged against the present only it is wrong because you are not worthier than your fellow human being. Thus you should contribute to society based on your ability of the moment, not rely on credits from the past. In the same vein, by achieving greatness at a given moment you will have no benefit in the future, because every moment is preceded and succeeded by a cleaning of yours and everybody else's slate. The absoluteness of the present is the ultimate expression of anarchic thought, because nothing you ever do will have temporal effects except at the exact moment you do it. It is popular to talk about living for the moment, and although, as was shown above, this is not wrong from the perspective of getting a taste of eternity, it is absolutely wrong and absolutely unworkable as a justice theory. The consequence is that justice must be also future orientated, and, if so, it must necessarily also be historically orientated, since future orientated justice will be a scam unless history shows that what was promised for the future will also be kept in the future. McTaggart's understanding of time in his prosaically named A Series, leading to the relevance of past, present and future, rather than just before and after, is thus perfectly relevant for the concept of justice. Justice does not need the before and after, the B series, but can only work based on past, present and future.

It may be argued that this is of little consequence for the issue of intergenerational justice, because our justice can be confined to us ourselves. Justice towards the unborn is not a necessary ingredient in all concepts of justice. In fact, in one sense, the whole debate about abortion is centred on whether there is a justice issue vis-à-vis the unborn. Yet, any concept of justice that negates the rights of future generations is based on a metaphysical model that few would subscribe to when faced with its consequences; the metaphysical model of après moi la deluge. Every person making a last will contradicts the model. Every person who believes that he or she loves beyond death contradicts it. Every religious person is necessarily adamantly opposed! A partial concept of justice is possible without looking to intergenerational justice, but a full theory must also consider this aspect. Which, of course, does not explain which rights new generations might have, perhaps, indeed, they might have none, but if one negates rights for new generations then it must be based on a considered and equilibrated model of rights, it cannot merely be après moi la deluge. Or, rather, in the case of immortality, no 'deluge', because there will only be 'moi'.

The realisation that future generations are potential carriers of rights is in itself an important ontological determination. Of course, this cannot translate into a right for every theoretical future life to be given life, as the number is almost limitless and the practical realities of childbirth and the hosting possibilities of the earth in any event constitute bottlenecks that cannot be overcome.[1] But there is a difference between accepting that not every theoretically possible life has a right of life and

[1] As the point here is all possible life, also the unconceived, the argument does not speak to the central theme of the abortion debate.

concluding that therefore all unborn life has no right of life. The negation of an individual right does not automatically lead to the negation of a collective right, no matter how individualist our societies have become. When it is argued that other species should not be made extinct there may be a self-interest argument in this, namely that we do not know whether that species might turn out to be very helpful for our species, but most people probably believe that there is a self-standing ethical obligation not to make other species extinct, which means that the species in question has a right to live. Not every member of the species, but the species as such. The ethical argument transcends utility and therefore encounters the usual suspicion that it is an argument of beautiful words, but no reality. Humankind, and nature at large, has a long history of destruction and the malaria mosquito would not be any more missed than polio. So why should we be concerned about the future of some obscure rain worm? And if we do not have to worry about the future of the rain worm, why should we worry about future human generations? Taking this line of argument to its logical conclusion we are back at the perennial question of why we should have ethical regard for other members of the human species at all, even the ones now living. 'No man is an island' is perhaps untrue, and perhaps we only do not kill other humans as a self-protection against being killed ourselves. 'We do justice that justice may be done to us' is the ultimate credo of utilitarian ethics. As we need not fear the rain worm or the unborn, 'justice' is not the issue. This would be the saddest of conclusions! We are in the end just atoms animated by atom interests; community interest is of no interest. Darwinian thought provides a strong overlay for this reasoning, albeit tempered by the overriding interest in passing on genes, but of course this logic is turned on its head if there is no passing on to be done because our immortality blocks. A Darwinian perspective without the overwhelming interest in gene transfer leads to the absoluteness of egotism, the only interest being self preservation, our lives not being a bridge to anything. Even the underpinning of love would disappear.

So is altruism fiction? In answering this it might be advisable to remember our attraction to moments when all things come together, because this is a time where our atomised existence seems to make sense as a part of a wider fabric. There is possibly an element of altruism in this. And it is perhaps particularly important to remember the role of the Other according to Lévinas. If we understand ourselves in relativity to Others then the Others are not only utilitarian objects, but integrated elements of our existence in their own right, and this makes it far easier to explain why we care about them. In certain situations it might be good business to become a mass murderer but perhaps you do not want to be understood as a mass murderer by Others, or, in fact, by yourself, since the Others are integrated elements of your own existence! This is the logic that fails when terrorists fly planes into skyscrapers, because there are suddenly two kinds Others, us and them, and perhaps even a paramount Other, God, in whose name you might believe you act.

Extending the perspective of the Others, it is part of the Darwinian logic that we mirror ourselves in our offspring, that highly significant Others are our children. For many, their children are the most firmly integrated elements of their existence. This might be argued to be Darwinian conditioning, but even if this were true it does not negate the logic that what we really want to do is to pass on our genes, or less

Darwinian, that a most important goal in life is to see the product of your love of your partner prosper and be happy. The pinnacle of love is to have a child with the loved one – sounds almost like a platitude – and it is questionable whether immortality could ever replace this most deeply felt purpose. Ultimately we might choose death in order to make room for our children. We might choose death out of compassion.

Perhaps the neighbouring question of ethical responsibility for other species, or, indeed, for our earth can be left aside at this occasion, suffice it to say that if the Other is the one who gives us reality then why should that Other only be human, why should our meaning not be defined also by other species and their diversity and by the earth and it many-splendoured nature. Many cried when Dresden was destroyed, not only because of the immense human suffering, but because a cultural wonder seemed to disappear. But Dresden, and all culture, is a mirror in which we can reflect ourselves; it is not utility in anything but the broadest of senses. We are naturally and inescapably anthropocentric, but that does not go to say that humanity is a self-standing structure of atoms, we are part of a larger fabric and we find our meaning in this belonging.

But intergenerational justice can also be seen in an almost juridical fashion: as an analysis of the basis of our possible claim to immortality and the resulting end to the generational flow. Many theorists believe that any claim is ultimately power-based. No power – no claim – no right. In a sense this is also what animates the idea of the absoluteness of the present. Do you have the power now, and if you do not then your entitlement is gone, no matter how virtuously you acted in the past. The living in the moment of power fallacy is sadly also what makes the anarchic dream falter. Power theorists might explain that the emphasis on power does not exclude minority rights, since they are ultimately a question of whether prevailing powers will allow them. Yet, is it not a terrible conclusion that justice is either divine or power based, it being noted that even divine justice to some extent is power based, because the threat is that unless you behave you will end up in Hell. Altruism again seems to get a bad rap!

Our power is based on being born and being the dominant species. But is that enough to deprive new generations of their power base? Can we just say with Stalin that the unborn should show how many divisions they have? Fundamentally, how did we get to be born?

Any power structure needs to be internally coherent to persist in the long run. Immortality is the ultimate long run, so a very high degree of internal consistency will have to be shown by any power base to ultimately succeed. Whether that will be the case is doubtful, as immortality might be thought to carry in it the seeds of its own destruction. With immortality we might believe that we have ultimately mastered our genes rather than them us. But the atomisation perspective leads eventually to the conclusion that our individual personalities are not the atomic building blocks, our genes are. Our personalities only dictate to our genes to a very limited degree, whereas our genes to a very large degree dictate to us. That is the biological explanation of the Freudian Todestrieb, discussed above. With immortality we will have silenced the drive of our genes towards our individual destruction, but clearly this does not mean that we have silenced the voice of our genes telling us that there is something more important than us, and that is the continual refreshment and recombination of genes. This tale is the tale that predicates evolution and it is hard to believe that its voice can be made to fall silent, even if in so many other ways we may be able to modify the building blocks of our existence. Stopping destruction is one thing, stopping the urge for renewal a completely different one. Geneticists might not have pinpointed what makes us, and all other species, so determined to pass on genes, but if and when that happens it is questionable whether the urge can be suppressed without making the whole genetic construction collapse. The divisions of the unborn are the armies of our genes that clamour for renewal, and as power bases go this would seem to be as strong as they get. Power may then ultimately be the A and O, but that does not make any given generation omnipotent. Does not mean that some kind of altruism has no power base!

Chapter 35
A Return to Immediacy?

That the lilies of the field and the birds of the air are always present in the moment of their lives was, as mentioned earlier, Kierkegaard's explanation of the immediacy of living which humans have largely lost. And it is, indeed, a characteristic of human beings to plan ahead and mostly be at some distance from the moments of their lives. Yet, it can be argued that there has been a degree of pendulum movement in this as well. The Greeks and Romans were for centuries unsurpassed in reflection and forward planning and the Renaissance and following centuries brought prosperity and the blooming of creativity and sophistication. However, in pre-Egyptian times, when humankind was emerging from the shadows, and perhaps in Europe after the Fall of the Western Roman Empire or during the plague and the Hundred Years War, the quest for survival was so immediate that the difference from the bird and the flower was much diminished.[1]

Modern times are a curious mix of sophistication, enormous ability to plan ahead, and impatience and craving for instant gratification, not only in material and sexual terms, but in getting every need filled quickly, even the need for information or knowledge.

This impatience to have every need filled quickly may obviously become highly detrimental in an immortality scenario, because we may fill the limited storage room of our lives much too fast. What is the point of quick and emotionally ungratifying sex if waiting for another couple of hundred years would bring something much, much better?

It is, however, possible that this confuses how immortality will make us react. The human ability to plan and create was brought about as the result of our fight for survival. Evolutionary mechanisms rewarded those with foresight. In an immortality scenario there will be little reason to reason or to plan ahead, since physical needs

[1] Nietzsche famously romantised the early Greek period of the tragedies as a period of Dionysian immediacy, thrown aside by the rationalism of the Socratian era. Realistically it must be said, however, that this appears to be a Nietzschian invention, suited for his purposes but historically and philosophically inaccurate.

P. Hulsroj, *What If We Don't Die?*, Springer Praxis Books,
DOI 10.1007/978-3-319-19093-8_35

will be taken care of automatically by a society of abundance, and, for all intents and purposes, reflections on death or its postponement or avoidance will have become unnecessary. Evolution will no longer be in play with humans.

So if self-preservation becomes automatic will humans over time lose interest in foresight and reasoning and ultimately lose the related abilities? Will this mean that we will return to an original state of immediacy akin to that of the bird and the flower? And will this be the solution to the boredom syndrome? From the perspective of the bird and the flower there might be little reason to fear endless repetition and everlastingness, because they are, as said, always present in the moment of their lives. With little reflection on the past and almost none on the future will humans return to the bliss of immediacy – and have found the source of everlasting satisfaction?

If this sounds far out it might be because we believe that future human existence will be, and should be, similar to current human existence, but why should that be so? Current urban existence has very little similarity with pre-Egyptian subsistence farming. If in the future we lose our sophistication is that not a return to nature à la Rousseau and hence to be welcomed?

Albeit counter-intuitive this solution cannot be readily dismissed. We might discuss human nature as hard-coded by evolution, but the reality is that over the last century we have said goodbye to quite a few traits that we might have considered impossible to remove from humanity. The reality is that one of the fortes of humankind is its malleability.

The basic thesis of this book is that our interest in immortality is an interest in continuation of our personality with its experiences and subjectivity. This you may argue would be fulfilled with a return to immediacy in immortality. After all, the bird of yesterday is the same bird as today, bar intervention by a cat, and equally so with humans in the future. We may lose quite a bit of what we see as defining us currently, our treasure of experiences and the ballast they give us as we face tomorrow's path. We would lose depth of personality, but this is a price we may be ready to pay, although not obviously so.

If, however, we scale up our perspective to that of the human species there will be those who would argue that the interest of the human species is equally to carry forward the collective identity and 'personality' – and that we have lost that if we accept rolling back the ability to plan, reflect and remember! In other words, what we could be ready to give up individually, if given a concrete choice, we might not be ready to give up for the collective future of humankind.

Chapter 36
Personal Relativity and Time

Unless we embark on the existence of the bird and the flower, immortality entails a degree of mental immobility. When a mind has lived through a lot, when it has a horizon of unlimited time, when it has nothing to fear, the mind will become increasingly static, as illustrated so nicely in the Makropulos Affair. Genetic manipulation will have achieved eternal physical youth for humans should they so desire, but how to keep the mind sharp and dynamic? The secret of immortality will be to stop cell decay or to ensure continued provision of strong, young cells, and this will be equally true for the brain as a physical phenomenon. So computing power and storage capacity might persist, but that is a different issue from how this computing power and storage capacity is used. Will storage capacity, in fact, have to be expanded in order to retain millennia of experience or will the mind's limitations mean that what happened 10,000 years ago would be hardly remembered whereas yesterday's impressions will live on for a few years. An inversion of how memory tends to decay with age, with short term memory going long before the long term memory. If ultimately the mind's storage capacity means that only, say, the last hundred years will be remembered is the person then the same as 10,000 years ago? Is only the infrastructure unchanged but the person ever new? In that case, we are close to the rebirth scenarios of Asian religions. Still, since it might be a rolling archive where continuity is ensured though the impressions of a new day being retained at the expense of the memory of an event long ago, one may assume that a certain continuity of personality will be present. This does not go to say, however, that the person of 10,000 years ago is the same as the person 10,000 years later, but this is, of course, also true for the person one day compared to the day after, as Hume has taught us. Only, with immortality the time proximities are lost and the personality fabric becomes much thinner. The human condition is that we gradually die and that we are gradually reborn, that every day brings us death and renewal, every day makes us forget a little who we were and every day makes us a new whole.[1]

[1] The movie Memento is an extreme variation on this theme.

P. Hulsroj, *What If We Don't Die?*, Springer Praxis Books,
DOI 10.1007/978-3-319-19093-8_36

Joseph Schumpeter has explained the processes of creative destruction in economics. The reality in our personal lives is that every day we are in the midst of a maelstrom of creative destruction, even if we do not perceive it as such. 'All changes, even the most longed for, have their melancholy; for what we leave behind us is a part of ourselves; we must die to one life before we can enter another' is how Anatole France explains it. With immortality time will allow, will dictate, ever more gradual personal reinvention; will remove us more and more from the anchors of our births and our youths.

?

I want to live:

- ☐ 100 years
- ☐ 1000 years
- ☐ 10.000 years
- ☐ 100.000 years
- ☐ 1.000.000.000 years
- ☐ Forever

Another way of looking at the field of tension between gradual death and continuous rebirth is to assume that with a limitless time horizon of immortality humans will learn to filter experience differently, will have to learn to remember differently. Perhaps human memory becomes meta-memory, where details are ignored and true substance retained. If this were so then immortal humans might start to become super-humans,[2] an embodiment of the Nietzschean Übermensch. From the wealth of information we will garner we will extract the essence and our judgments and conclusions will be so much better founded. Our memories will be even more synthetic, more vertically integrated than now. In this hopeful scenario human lives will still be dynamic, and we will only move from the equivalent of the hectic of the day-fly to the majestic calm of the elephant or the giant tortoise. How such a sublimation of our capabilities would come about is unclear, however. The more likely scenario is that as humans get older and older they would become more and more cantankerous, more and more annoyed at seeing yet the same thing again and again, almost like the worker at the assembly line who every day has a Groundhog Day experience. The truth of the biblical saying: "What has been will be again, What has been done shall be done again, There is nothing new under the Sun" would be brought home poignantly.[3] The arc of development and decay that now characterises human life might largely stop or become so emasculated that the interest with which we observe our own progress and regress might almost vanish. So little progress, so little regress to observe![4]

It goes with the territory of Lévinas and the Other that there might be limited attraction to just watching one's own navel, be it in the compression of current time or the immensity of immortality. It might also be said to be of limited interest to interact with persons who remain largely unchanged through eternity as a result of the unlimited gestation time that immortality will offer. Even if the limits of human development were expanded in eternal time clearly every nugget of development would be rare and when one materialised it would be awfully likely that we would have seen that specific development before. The essence of human relativity as we know it is a dynamic element with both observer and observed: we change while we see other change. If we no longer change at any significant speed, and others do not either, an idleness element is introduced which at least with our current conditioning we would have difficulties handling. Nietzsche and eternal repetition comes to mind again, eternal boredom comes to mind. The pride and interest we take in our own development and that of our loved ones would be hard to sustain.

Some human beings seem to be temperamentally very accurately attuned to the speed of their lives, others seem to be always ahead or always behind. When aver-

[2] A bit in this direction Christine Overall in David Benetar, Life, Death and Meaning.

[3] Ecclesiastes 1.9.

[4] It has been suggested that with the promise of immortality, but still the risk of death, humans would become more risk averse. This is debatable. At the start of life perhaps, but after a billion years hardly, since the marginal attraction of further life and further boredom would seem to dictate a much higher risk tolerance, see Christine Overall, footnote 2 above, on this sort of discussion.

age lifetimes were thirty it was hard to be ahead of the speed of life, but as life expectancy increased it became easier, and it became harder to be behind. With longer lifetimes impatience grows. Assume that puberty is slowed to set in only with the age of 100 or 500. It would be hard to conceive of a very satisfactory psychological situation for humans, since our mental construct is aligned with a much faster development process. Just as our memory might not be as malleable as it would need to be for immortality to become a real blessing, our impatience might be too hardcoded as well. Even the miracle of genetics might not be able to help, and even if it could, would we want it to? If the sum of human experience and feeling were more or less the same regardless of the lifetime involved – if there is an overflow point in every human being's capacity to absorb – would we really want to prolong the lifetime to eternity with the result that we would be sitting around essentially watching the paint dry?

Chapter 37
A World of Immortals

The morality of immortality can be understood as being an issue of individual conscience: when immortality becomes possible will I avail myself of it? And it is certainly true that there is a question of individual morality involved. Yet, the ultimate question is societal and requires a societal response. When immortality becomes possible will we as a society allow populations to avail themselves of eternity? This is, indeed, a much more complex debate because it juxtapositions our cherished individual liberty with broader ethical considerations; considerations that might lead to unwanted and avoidable individual death in order to satisfy collective interests, such as maintaining a viable community and safeguarding the rights of future generations.

A fundamental question is going to be whether a society of immortals will become more virtuous or more depraved. With longer life times and more wealth we tend to believe that our society has become more virtuous, which is a truth with modification when we consider the millions of avoidable deaths we allow every year through ignorance and neglect. Still, there is an observable dynamic in this which we might hope would be taken to its extreme in a society that would not only be immortal but also rich. We might hope that the tension between anarchy and society will leave society the winner, since the immortal society is only a blown up version of our current society where, indeed, community has taken hold. However, immortality introduces two factors not present in our current societal bargaining, that is, the enormity of time, and, more importantly, the absence of death and the resulting elimination of fear of divine retribution for depravity in earthly life. To be fair, with the increased irreligious nature of many communities there seems to be no anarchic effect, some of the most irreligious are some of the most societally cohesive, take Scandinavia. In contrast, the radicalisation of religion in many regions of the world has led to more cruelty in the pursuit of some alleged higher goal. As we know from history as well, the prospect of eternal damnation or eternal reward can lead to the most atrocious acts. And that is also the rub in terms of the immensity of time that immortals would be faced with. Some religious fanatics will do whatever they think will bring them eternal reward and immortals might deploy

P. Hulsroj, *What If We Don't Die?*, Springer Praxis Books,
DOI 10.1007/978-3-319-19093-8_37

the same logic in order to achieve eternal terrestrial reward. Perhaps in Scandinavia you are tolerant because you know that life is short and cannot be expected to be followed by anything, so there might not be much point to evil, since you will only enjoy its fruits so briefly. Why spend life fighting for the ephemeral, instead of just squeezing out the maximum of what you are legitimately entitled to, and that you do not have to spend too big an effort to defend? Conversely, would we not be ready to commit the usual atrocities that we do to reap the rewards of heaven, if, after immortality becomes possible, heaven will ultimately have to be on Earth? Slaughtering a few immortal neighbours in order for you and your kin to have much greater comfort for eternity might start to look suspiciously attractive in our so pervasive 'end justifying the means' logic. A very likely ultimate result would be that, indeed, villainy would become pervasive, both because of 'the end justifying the means' and because of the boredom effect that will mean that, after trading-off for 500 years whether some heinous act could be justified, in many instances surely temptation would win out. Navel gazing of any sort has rarely led to much good and endless such will likely lead to even less.

Drawing the societal consequences from this might bring the conclusion that villainy will become the droplet of toxin that will ultimately poison the whole well, because villainy and defence by the righteous will play out in eternal time. Villainy might make the whole immortality edifice collapse under its own weight, as warring factions will form and turn the potentiality of ideal society into actual societal nightmare. A society embarking on immortality must consider carefully whether societal tools will be available to master the genie once it is out of the bottle.

Chapter 38
Natural Selection and Immortality

When ultimately society holds the trigger on immortality, it will have to decide how to exercise this power. It is, of course, a safe bet that elites, political, financial, cultural, will be the first to profit, no matter the rational selection criteria. But this cannot mask that society is likely to strive for perfection in the chosen ones if immortality cannot be made available, or will not be made available, to all from one day to the other. The effect is that eugenics-like considerations will raise their ugly head, but also that an amusing issue will arise, namely how to define perfection, and whether we really want it for ourselves and for others?

It would seem to go with the territory of immortality and eternity that the curing of all human infelicities would be the logically attendant ambition. And thus perhaps we will find ways to modify our genetic coding while we live, which might be satisfying in the optic of eternity, but would put into question again what sort of continuity must be present for individual identity to be retained. If it becomes possible to go from klutz to Federer-like ability as suddenly as, medically assisted, the flat-chested can now become buxom, it is somewhat hard to detect the continuity of existence. But, at least, society will be saved the horror of having to eliminate you because your genetic make-up is less-than-perfect.

If we do not learn to repair our genetic imperfections as we go along then society will have to choose between respect for the equal value of each human life, with the consequence of eternal imperfection and the path blocked for more perfect members of the next generation, and the eugenics route of blessing only those who are already blessed. Brave New World would appear a romantic comedy compared to the latter! Onto that comes the boredom element. If society chooses universal perfection it will in some way or other churn out perfect samples according to a commonly held notion of perfection. The idea that one would spend an eternity with individuals who are perfect and not only similar among themselves but also with you is again a Bokanovsky Group nightmare of physical and intellectual disstimulation. Sex between clones is a truly unappetising idea.

The soundest choice in terms of genetic variation between humans would be to strive for as much complementarity as possible. But in the final analysis any

P. Hulsroj, *What If We Don't Die?*, Springer Praxis Books,
DOI 10.1007/978-3-319-19093-8_38

'planned economy'-like system of genetic attribution and variation, even against the backdrop of eternity, will be unethical and would seem to overestimate the human capability to choose wisely. The human being always seeks to eliminate the random. But perhaps the random is essential for sustained happiness. Leaning again on Wittgenstein it might be argued that happiness would not exist without the existence of misery, that the deliberate might have no meaning unless the random exists as the clay out of which the deliberate takes it shape.

In the absence of the random a world of immortals could then, as mentioned, become a boring world of perfect individuals, hardly discernable one from the other. There is an even darker way of looking at this, however. Depending on elite strength, elite tactics and perception of elite interest, a two-class society could arise, where an Epsilon underclass in classical Brave New World fashion is bred to serve the perfect Alpha upper class. It is perhaps doubtful whether such an underclass would be necessary in view of the prospect of all-pervasive automation, but if it were there would be, beyond the ethical question, a power retention dilemma. If one assumes that the new underclass also consisted of immortals we would be talking of eternal servitude, something hard to imagine as sustainable even if, as in Huxley, the underclass was bred to be simple-minded or be made into chimeras. Foucault's biopower, the power of the state to control our physical condition, might have reached its limits. If the upper class was immortal and the underclass mortal it is safe to say that the upper class would find itself to be more mortal than hoped for, as the underclass would surely rise. And if the underclass were then given reluctant immortality we would end in a classical Catch-22 Whatever you did as elite would lead to a restless underclass.

The conclusion must be that a world of immortals can only be a world where all humans are immortal and politically equal to the largest extent. This might be a hopeful message in a political sense, but perhaps only in that. The apple of knowledge would be likely to mean that we would be expelled from Paradise a second time; this time from the one we had tasked ourselves to create – but could not, because both human perfection and imperfection would be catalysts for destruction.

In any event it is too facile to just state that because all humans are created equal then also all humans must be allowed to seize immortality. There would be terrible eugenics-inspired questions on whether, indeed, lives lived with debilitating illnesses should be perpetuated when the sufferer is not able to choose himself. And if human personality did not become completely malleable, society would have to decide whether it would allow an eternity of criminal behaviour or whether immortality could be forfeited. The Viking god Loki demonstrated how badly things can turn out when evil is given divine time, and when gods possess the potential of immortality and the potential of death. Exactly this would be the position of the immortal humans. Immortality would be reachable but would be no certainty. When would the privilege be withdrawn?

When humanity achieves the power of immortality it will have put itself in the position of god. Even if god exists he will lose relevance for those who trust in earthly immortality. God may be there, but will never be seen. In contrast to all monotheistic concepts humans would have created an extreme form of polytheism, where every human is a god. Democratisation of divine power may sound good, but what monotheism achieved, as opposed to Valhalla, was that omnipotence would be exercised in a necessarily unitary fashion. With billions of gods how would it work? Would good and bad become irrelevant; replaced merely by useful or not useful? And useful for all or useful only for me, the immortal? Would the spiritual, so closely linked to the mortal, become less relevant – or more? Would immortality be the path by which humanity reverts to original matter, after curiosity, striving, joy and virtue have become more and more irrelevant? Is death the ultimate animator?

When humans are in a position to not only exercise limited power, but thanks to genetics and technology almost unlimited power, will each of us, strong or weak, be ready to be god? Would god choose to be god, if he was not god already?

Chapter 39
The Tragedy of the Longest Life!

The assumption in this book is that immortality is a choice and hence that immortality is not absolute. How the individual can deal with this freedom is the subject of a future chapter. There is another aspect of importance, however, and that is how the loved ones of an immortal will exercise their freedom to live or die, and the effect of such extraneous choices on the immortal. This issue is addressed here.

When immortality becomes optional, some may accept the offer, some may not. A question of individual freedom, it may be argued, and surely this is basically correct. Yet, human beings are islands in the stream, so even if the immortal has gained independence from death he or she has not gained independence from others, neither from friend nor from foe. In a sense, the reward of immortality can be understood to be the ability to continue human relationships beyond the current horizon of our lives. But herein lies the rub, as well. If you choose immortality it does not follow that everybody you love does the same. Immortality might become a mixed blessing if those you love decide that they have had their fill long before you, as will always be the case if your remain immortal. Imagine lovers parted by the chosen death of one, perhaps after 10,000 years, but still leaving the partner to face everlasting grief.[1] The survivor might then, of course, choose death despite

[1] Wordsworth, Intimations of Immortality, might then gain even more poignancy:

'What though the radiance
which was once so bright
Be now for ever taken from my sight,
Though nothing can bring back the hour
Of splendour in the grass,
of glory in the flower,
We will grieve not, rather find
Strength in what remains behind;
In the primal sympathy
Which having been must ever be;

P. Hulsroj, *What If We Don't Die?*, Springer Praxis Books,
DOI 10.1007/978-3-319-19093-8_39

opposite inclination, but this then only goes to show that immortality has a deep shadow-side beyond just the boredom aspect. Immortality might buy freedom, but not everlasting love!

It has been suggested that in earlier centuries parents did not love their newborns or young children with the same depth of emotion as is now commonplace. It is said that the reason was the high mortality rates of infants and young children. Parents only started full emotional investment when, with age, the chances of survival of the child had increased significantly and with that the chances of a good return on the emotional investment. This being so, the immortal could then perhaps be assumed to be reluctant to invest emotionally in non-mortals,[2] be they lovers, family or friends. Immortals would perhaps create peer-societies, because with shared endlessness even small emotional investments would become exceedingly valuable over eternal time. A dollar invested well in 1800 would yield 114 dollars of compound interest in 2000. Same perhaps with emotional investment – imagine the effect of the compounding factor over eternity! Although this may be an alluring perspective, the question remains, however, how interesting anybody can be as a social partner over the billions of years. Perhaps the interest will not compound, but rather the capital will decay. After 10.000 years friend might have become foe, which might still be better than the ennui that might set in after 30,000 years when former friend, turned foe, might have become an incredible bore.

The logic is hence probably the inverse. In the world of the immortal every new stimulant, every new friend, every new loved one will be particularly prized, because, at least, there is novelty in an ocean of the well-known and already experienced. The investment in the new, the temporary, is likely to be very high. Any unique moment can be expected to be highly emotionally treasured – as much as Makropulosian coldness will allow. The immortals are unlike to form ghettos, they are likely to try to become poles of attraction for the transient.

To the extent it can be surmised, those who have chosen life-times will be in demand, particularly if death is rewarded with the right to procreate. The immortals would have an interest in creating beehives, with immortals as the queen-like fixtures and the mortals as generations of worker bees. The immortals might ultimately become god-like, except for the fact that the mortals will have the greatest bargaining power, because the immortals need them more than they need the immortals. Family might still be important, but as networked hives, rather than just one shared by the immortal family in suffocating closeness and endlessness.

The revolting feature of immortality is that death of loved ones might in a roundabout fashion turn out to be welcomed in the midst of the emotional wasteland.

In the soothing thoughts that spring
Out of human suffering;
In the faith that looks through death,
In years that bring the philosophic mind.'

An overly hopeful perspective, it would seem.

[2] See Christine Overall, in David Benetar, Life, Death and Meaning.

For an Elena Makropulos the feeling of loss may ultimately become the deepest feeling possible. Immortals might thus become akin to current humans who treasure tragedy, perhaps unwillingly. For immortals losing what they do not want to lose might be more fulfilling than any desired achievement – because they would expend exhaustible capital – because of the attraction of opposing sentiments. The power and determinism of impotence.

Eventually the ultimate attraction for the immortal might be their own deaths, because this is the only thing they did not experience in their endless lives, and the heat of the wanted rubbing against the unwanted might bring a final Faustian moment of the most contradictory form of happiness – the happy death as the result of the last remaining, untried experience. No more pull of life, perfect stillness in the surrender, ex hoc momento pendet aeternitas, the eternity of Selbstaufgabe. Perhaps a Kierkegaardian leap into eternal rest, eternal perdition – the ultimate goal of Hinduism and Buddhism having been achieved!

Chapter 40
Choosing Life!

Whether humans will ever gain the ability to call something or somebody to life might well be doubted, as, perhaps, overstressed above. However, the first principle of epistemology might even surprise us in this domain, and if so, humans would be faced with yet another host of difficult issues which, in any event, are relevant subjects of study.

Walt Disney might from his permafrost abode hope that a future generation will choose to call him back to life. Ontologically we can ask whether really it would be Uncle Walt we would raise from the dead, or only his vehicle. Would Walt Disney's memories and identity be resurrected if humans could give renewed life to his well-preserved physique? If not, what would be the point? The issue is the same as for cloning: will the same DNA make the same man? Unless personality and memories can be re-awakened there does not appear to be a lot of reason to re-create earlier life. At least those specific DNA combinations already had a bite of the cherry, so giving them two would appear unfair, when the pipeline of future generations would be so terribly backed up. That having been said, it can be argued that there would be a strong Darwin-in-reverse motivation for calling back to life earlier generations, through cloning or unfreezing of the permafrosters. A descendant can be argued to have almost the same reason for reawakening parents as for passing on genes to a future generation. In both instances it might only be the vessel that is created – former memories and experiences must be assumed to be gone from the ascendants. But even so will the loving child not want to give life to the Ebenbild of a loved parent, giving the parent a highly unusual opportunity to both have served as parent and becoming the child of the own child? Darwin taught us that we strive to pass on as many as possible of our genes. With parent rebirth a proven and loved DNA combination would be recreated, and although that might not be good for evolution, it could perhaps be believed that it would be good for love. First you would clone yourself a few times and then your loved ones a few times. An eternity would be created where not only eternal repetition of events would loom large, even life itself would be repeated in a cloning cycle that would ensure the comfort of the presence of the vessels of loved ones whilst navigating the sea of never-ending repetition.

P. Hulsroj, *What If We Don't Die?*, Springer Praxis Books,
DOI 10.1007/978-3-319-19093-8_40

Is that really Uncle Walt?

As you never fully share the personality and experience of an Other, although you find meaning through the Other, the special seduction of a cloned parent is that the parent him- or herself might 'regret' having lost the past, but for you, as 'user' of the Other, it is not so important that the common past has been lost since you would still reflect yourself in an Other with all the inherited characteristics that you have learned to love![1] DNA reproduction would allow bi-directional creation, new lines of descendants and recreation of long lines of ancestors!

If it is assumed that earth will only be able to host a limited number of individuals the possibility of bi-directional creation of life will add immensely to the tension not only within society, but within each individual. If you are given the possibility to create only one new life, do you choose parent as child or child with your partner? One can assume heated discussions between partners around the dinner table: so you prefer your father to our common child!

It is, of course, possible that colonisation of the universe will mean that the dilemma of which life to choose will not become too stark. Perhaps humankind will not be limited to our little earth but will find ways to expand into the universe in such a fashion that the dilemma of choice will only materialise after the passing of many millennia. Still, even such new habitats would not be a panacea. Different hosting planets will offer different living conditions, some much more pleasant than others and, with humanity growing and spreading, species homogeneity will also be under pressure. Fratricidal risks will go up.

With earth as the possible only place of human settlement, and therefore Malthusian constraints being present, every new life will become immensely precious. But, even disregarding ancestor revival, how will society choose – and can a society enforce its power of choice? China might have been successful with its one child policy, but the game will be played differently when eternity is the backdrop. Those who control the bottle of eternity, the technology enabling immortality, will possess tremendous power if the technology is not in the hands of everyone, and how could such an imbalance persist? If it is in hands of everyone anarchy will rule as everyone will choose their preferred new life. Absolutism would not even be a conceptual possibility as absolutism will never be able to eternalise! The colonisation pressure would increase tremendously, as we know from general human history, and with that the inducement to cut ties to a civilisation that turned out to be less accommodating than desired. The prospect of different human species located in different parts of the universe might be the path to which immortality will lead us, and in this sense immortality is perhaps only one step in evolution, an evolution that will allow new worlds to arise! Perhaps human life on earth will turn out just to be the seed for universal life, guided not only by natural selection but the conscious choices of the subjects of evolution!

[1] Nathalie Cole singing a digitally enabled duet with her father Nat King Cole and the music video in which Tupac's hologram stars are steps in this direction. Surely, digitalisation will soon allow us to emulate not only Elvis's voice but allow the creation of his alternate hologram existence, so that we shall be the beneficiaries of endless new Elvis hits, full of soul of the most ultimately soulless character. Holograms might, of course, also handily complement future machine based lives, giving our digitalized souls a semblance of physical appearance, a physical appearance much more convenient than our bodies, with holograms being so much more readily repairable.

Chapter 41
Humans and Humankind

Having children is at the present time the only avenue for humans to secure some sort of earthly immortality. We feel that we pass on part of our personality, part of our consciousness even. This is the propaganda that our genes so realistically succeed with, although from the gene perspective the totality matters little, and consciousness even less; the passing on of the individual gene is the driving force. Humans believe they are totalities, but in the quest for immortality they are, in some respects, atomised. And yet the game is to bring across as many individual genes as possible, so as vessels of a multitude of genes the totality still matters. If we could pass on all our genes this should be our inclination according to evolutionary theory – something to remember when discussing cloning and the like – something to remember when discussing love, because it can be argued that both love of partner and love of children are predicated on gene survival. But if no others are needed for the gene game, why engage, why love; an activity that then might appear entirely wasteful, and hence against Nature's design in the future?

It can well be asked how well immortality works for genes. How many genes have survived from the start of organic life; how many from the early days of living beings; how many from human erectus and the time of the passage from Africa? The answer is that immortality works well. All humans consist of many genes shared with early humans, many with pre-humans and quite a few with early organic life. Yet seen from the perspective of each vessel, each human, the answer is not quite as rosy. What we are interested in is not just the survival of the species, but the survival of elements of our specificity. So a human is interested in the survival of his or her combinations of genes, and, in rare cases, of a mutated gene unique to that specific human. Blue-eyed persons may be interested in seeing future generations blue-eyed, but, of course, there is no unique contribution from an individual to future generations in this sense, as there are many blue-eyed persons sharing the dream. Unless a person is the possessor of a unique, mutated gene, it all becomes a numbers game. If you as a blue-eyed person pass on the blue-eyed gene-combination you add to the mass of blue-eyed humans – like in a democracy you have added your blue-eyed vote! The uniqueness of you is how your genes are combined, but in

P. Hulsroj, *What If We Don't Die?*, Springer Praxis Books,
DOI 10.1007/978-3-319-19093-8_41

ten generations your uniqueness might be highly diluted, in a hundredth normally beyond any recognition. Individual genes might well have survived – might well have a provenance directly from you, your blood-line might not have become extinct, but does that make you immortal when your contribution has become so insubstantial and so un-unique? If your genes were only combinations of genes shared with many others, the genetic make-up of your successors ten generations hence would be unlikely to carry your imprint very strongly. And then there is the business of the many bloodlines becoming extinct over time!

In sum, the genes are lying to you: they bring comfort beyond death, but they surely do not bring any discernable immortality. It is not as easy or as cool to become Abraham as we might think!

> Your children are not your children.
> They are the sons and daughters of Life's longing for itself.
> They come through you but not from you,
> And though they are with you yet they belong not to you.
> You may give them your love but not your thoughts,
> For they have their own thoughts.
> You may house their bodies but not their souls,
> For their souls dwell in the house of tomorrow, which you cannot visit, not even in your dreams.
> You may strive to be like them, but seek not to make them like you.
> For life goes not backward nor tarries with yesterday.
> You are the bows from which your children as living arrows are sent forth.[1]

If we individually, and our genes, are then not immortal there might be appeal in the thought that, at least, humankind is immortal. Our specific genes might not possess Abrahamic qualities, but our collective treasure of genes might. The procession of generations might give a consoling feeling of continuity. And, indeed, continuity there is, but immortality hardly, unless humans can truly master nature in the fashion mooted in this book. Immortality of the species will most likely be unattainable in the normal evolutionary scheme of things, because even as a species we are fragile. Humankind may, indeed, be around for many, many generations, but we are unlikely to have the durability of many strains of bacteria, simply because higher order organisms face more subtle challenges than simpler ones, due to their more evolved state. Dinosaurs are the trivialised example, but it could be argued that the more fine-tuned a species is relative to its environment, and that is a feature of evolution, the more susceptible it becomes to changes in that environment. Against this it may be argued that humans are a demonstration of exactly the opposite. Humans have learned to master every environment, even beyond earth, and have learned to dominate earth rather than be dominated by it. The use of tools, and language, has been our path towards environment independence. Many humans believe that this has made us all-powerful, and moved us beyond the realm of dependency. This is both right and wrong. It is true that we have gained immense autonomy from Nature, but it is equally true that through this autonomy we have created intra-species dependencies by pursuing so vigorously the logic of the apple of knowledge,

[1] Khalil Gibran, The Prophet.

think nuclear armament. Our high sophistication makes us susceptible to highly sophisticated threats created or fostered by us ourselves. Over the centuries or millennia humans will likely prepare their own destruction in one form or the other.

The rewards of high emotional and intellectual capability are that we get an ever more refined possibility of grasping glimpses of eternity, ex hoc momento pendet aeternitas. The price we pay is our ultimate perdition, even as a species.

Chapter 42
Immortality for Humanity, Darwin for Everybody Else?

With chimpanzees sharing 99.9 % of genes with humans, speciesism should already be a concern high on the human ethical agenda. Imagine now that progress in genetic engineering enables the introduction of another 0.05 % of human genes in the chimpanzee. Would the result be an animal, a chimera or some kind of human being? This debate, which seems highly likely to face humanity in relation to a number of other species given current abilities to implant extraneous genes, will, of course, in an immortality context, raise the question whether immortality is reserved only for 'pure' humans. Same thing the other way round, by the way. When will the human genetic make-up be so depleted by introduction of animal genes that the resulting being will no longer be human?

Clearly, there are concerns here that we might have great difficulties confronting, and these concerns will be compounded by the question of whether the gift of immortality should only be bestowed on humans. The gift itself could logically also be applied to other beings, since we would be masters of the process of aging, but would we, should we, share?

When making ourselves our own gods we must also decide whether humans will want to be only gods for humans, or gods for most living beings – as, in many respects, we already are. Only, possible animal immortality raises that game very significantly.

Before turning to the ethics, consider the economics. For instance, having to raise animals for milk production is a costly and wasteful exercise. Would it not be so much simpler to make the dairy cow immortal? As usual the Vikings were ahead of the curve. The gods in Valhalla had the pig Saerimner which every evening would provide meat for hungry gods, only to have its sides re-grown immediately. Much in the same fashion immortality could make the well-proven dairy cow an eternal fixture. Much loved pets could become eternal companions (but see if even that love could survive the pressure of endless time). As we perfect flora and fauna why not opt for immortality for our creation? Why repeat the mistake of our god to make death part of the equation?

P. Hulsroj, *What If We Don't Die?*, Springer Praxis Books,
DOI 10.1007/978-3-319-19093-8_42

Saerimmer forever!

The ethics issues go beyond defining what human is, and what is hence, eo ipso, entitled to immortality, and include issues such as whether, in the Jain fashion, there is an obligation to eliminate as much death and as much pain as possible. Would it be correct to let animals face the unpleasantness of death, if we ourselves have escaped and could spare many animals the experience (which does not go to say that veganism or vegetarianism is a moral imposition, but would imply that due care is)? After all, when a certain animal has ongoing utility, pain could be reduced by making it immortal.

The difference in approach to death between humans and animals might be that while animals seek to avoid death, the concept of death does not animate animal life in the same fashion as it does human life. It might be assumed that animals almost by necessity live for the moment (viz. Kierkegaard), whereas humans tend to live with a much broader perspective on time, and therefore on death. A binary approach to life and death – yes, I live, now I die – might be understood to be more in harmony with the singular focus on renewal through the passing on of genes than the human sophistication of fearing death; yet, in the final analysis that is just another way of saying that human personality and human processing of experience is more developed than the equivalent in animals. It is highly debatable whether this difference in degree of 'evolution' is ethically truly relevant.

In other words, it could well be asked whether more highly developed species than humankind, which the universe might contain, would recognise an ontological distinction between human and chimpanzee. If humans just happen to be the masters of the game on earth must we not, in fairness, share our accomplishments as much as we can?

The idea may appear abhorrent, primarily because we in this example would start to harbour almost pantheistic feelings of how we should not interfere in nature's work. But, of course, we do, and therefore in its final analysis the issue becomes one of why we would not intervene in nature's work when it comes to animals, but would when it comes to humans. Clearly any species has a certain bias towards its own interests, and that is not to be condemned, but it would not appear to be so clear that nature should not be allowed to take its course with us, if it is allowed to take its course with our fellow living beings. If we reject immortality for animals, perhaps we should reject it for ourselves, as well!

Chapter 43
Gaia and Dystopia

The assumption in earthly immortality is that humankind will become the master of life and death. Yet, as also explained, there might be features related to human nature that will make it impossible to enjoy the status of immortality. We might be able to find the seeds of our own destruction in ourselves. But we might also be able to find those seeds elsewhere.

James Lovelock has been much derided for his suggestion that earth and universe are kept in balance by the interaction of forces we only barely recognise, let alone understand. And although the proposition remains unsubstantiated and controversial, the Gaia label has entered the general vocabulary and its central hypothesis informs many discussions.

In the context of immortality a transposition of the concept could give comfort to some that immortality will never become an issue, because nature will re-establish balance even if humankind had the technical means of stopping aging. However, such a belief would challenge our Aristotelian faith in scientific method and scientific progress.

In good scientific fashion we always analyse causality and where we can detect no conceivable causality between a proposed way of proceeding and suggested negative effects we believe that everything is alright. Yet one of the interesting features of statistical method, exploited by freak economics, is that it can demonstrate correlation between factors where there is no obvious causal link. A twisted way of applying this method might then allow us to extract new meaning from some of the disasters that have hit humankind in the past. Might show correlation between radically changed human behaviour and highly adverse reactions by nature.

Black Death followed intense city settlement activity by humans, and promptly decimated populations by up to half. We now know of the deadly influence of the rat flea, but in the end the rat flea was given a habitat by changes in human habits and habitation which, in a sense, had the effect of restoring a previous balance, Gaia-like, if only for a little while. The Spanish flu was very possibly also a reaction to unusual human behavioural patterns associated with the First World War. In killing the young and strong more than the old and weak, as normal flu does, the Spanish

P. Hulsroj, *What If We Don't Die?*, Springer Praxis Books,
DOI 10.1007/978-3-319-19093-8_43

flu relied exactly on the strength of the immune system to create a cytokine storm, an immune system overreaction. And the amassment of young, strong soldiers in garrisons and at the front made transmission easy. So the Spanish flu was a way for nature to exploit human weakness in radically changed circumstances. The HIV/AIDS epidemic was clearly enabled by the radically changed sexual mores of the flower power generation, and nature was not very accommodating.

The upshot of such examples is that we have learned the very hard way that radical change of behaviour carries an inherent epidemiological risk. Still, we have comfortingly re-established causality, and immortality enthusiasts can point to how epidemics come and go, but how they do not stop the human project. The plague did not stop human settlement, the Spanish flu did not stop large scale war, and HIV/AIDS not sexual freedom. The argument can then run that although the path to immortality might not be smooth, radically changed behaviour will always find a way to succeed.

There is a scaling problem in that proposition, however. With an immensely radical change, such as immortality, nature will be given a very broad staging area for its attacks on our putative strength, and although we might hope to be able to defeat each attack one after the other, it is not obvious that the consolidated strength of the attack can be ferreted out in such a way that each challenge can be separately met. But even if this was the case, there may be, in the Gaia way, interrelationships within our physical world that will mean that the attack will not only be on our immortal health. Some may argue that climate change is one such attack. Others may argue that this line of argument is just another step in the direction of pantheism.

In the 1970s a much discussed book was The Limits of Growth. Its basic thesis was eventually disproven, like that of Malthus, and yet works like these do raise the valid issue of whether human expansion, and human ingenuity, will at some time meet hard boundaries. This is a question that, of course, is put at its epitome by immortality. And it is a question that ultimately is one of whether humans are tied to their existence by a number of almost subliminal bonds, which cannot be broken, or which can only be broken at the price of losing humanity, an example of the latter perhaps being machine-based life. The limits can be understood as physical, say, epidemiological, or psychological, of which the psychological limits might be the more difficult to grasp. There is much romantic talk of the ties of given individuals to terroir or a certain culture, with these individuals withering away when removed from their nurturing environment, and although this lesson cannot be generalised there is a relevant immortality issue in terms of whether humans in their fundamental genetic constitution are Totgeweiht, meaning that humans must feed off death, and whether this remains true even if we learn to stop the aging process of the genes. Stopping aging might be one thing, ability to live forever another!

Chapter 44
Beyond the Limits of Rationality

Two thousand years of Socrates, Plato and Aristotle and 300 years of Enlightenment have given us great faith in the benefits of rational reasoning. And, indeed, the benefits have been huge. The prosperity and peace that a significant part of the globe experiences are largely the results of rationality. However, looking at human existence though the prism of immortality puts so much doubt into rational inquiry that one can be tempted to think that all the issues related to cogito ergo sum are somewhat trivial in comparison. Cogito ergo sum elevated the study of subjectivity to new heights, but immortality adds to the subjectivity dimension so many objectivised questions on what human existence is about that the boundaries of rational and subjective discourse become blurred. The issue before us is whether phenomena such as machine-based human life are really human, desirable, avoidable, sustainable; whether earthly immortality can lead to liveable lives; whether limitations in the human capacity to feel and remember would render immortality inhuman; whether the elimination of the creation of new generations can be justified or would make sense. In the earlier discussions many of these themes were attempted analysed by the use of rational argument, but this does not allow us to escape the fundamental question of whether these issues ultimately transcend rationality. In this respect immortality might raise issues similar to quantum mechanics, where the currently prevailing argument seems to be in favour of a law of no-law. If the laws of physics and predictability do not apply fully in quantum mechanics, perhaps the laws of logical reasoning do not apply fully to human existence and immortality?

There are two ways of understanding limits to rationality, one being that the topic cannot be explored by means of logic, and the other, that human knowledge is still so imperfect that it cannot serve as a foundation for rational discourse on a given topic, even if, with time, that may become possible. Starting with the latter point, it is clear that humans must deal with a large number of issues where knowledge is more limited than it could be, yet decisions and conclusions must still be made. Sometimes we label such decision-making as 'common sense', but, truth be told, the history of common sense decision making is less than glorious. Still, common sense is rather hard-coded in us, sometimes perhaps as a practical way of synthesising

P. Hulsroj, *What If We Don't Die?*, Springer Praxis Books,
DOI 10.1007/978-3-319-19093-8_44

diverse information into an actionable format, but often also as a way of overcoming our knowledge shortcomings. Induction might obviously sometimes be a palatable method for overcoming information deficits, but since reality is often wilder and more unpredictable than we believe, common sense based on induction has been prone to lead us into the wilderness. Plato's shadows are perhaps more difficult to give meaning to than even Plato assumed.

Analysing immortality certainly has much to do with trying to overcome information deficits, since humans do not have experience with life over extended periods.[1] Average human lifetimes might have more than doubled in the last hundred years, but that does not truly help us assess increases by magnitudes, let alone eternity. Other living beings have longer lifetimes than humans, but what can we extrapolate from the lives of turtles, other than the utility of a thick shell? A classical way of overcoming this dilemma could be to postpone the issue until such time when more information becomes available. It could be suggested that we should not start to worry before human lifetimes have increased to 200 or 300 years. As a matter of fact this might not be the worst idea, except that once on the path to 200 or 300 years will the process be stoppable? Immortality might be upon us without us truly considering whether we want it. In the end the evolution might be so fast as to be more like a revolution, leaving humankind, as so many times before, to make decisions after the fact.

Ultimately it can be argued that these issues on information deficits and the imperfection of decision-making on that basis take second place to the more fundamental problem of whether immortality can be rationally dealt with, no matter the amount of information. There is still a good chance that more information will allow us to solve the mystery of quantum mechanics, but the same might not hold true for our understanding of the real nature of existence and humanity. And despite the admonitions of Kant's Critique of Pure Reason we might not be able to escape making judgments. Transcendence might remain a mostly closed book, but we will still have to make existential judgments not only at the individual but at the societal level.

The reason the real nature of existence and humanity might defy rational analysis is not so much that the human mind can not grasp the complexities, but more that the intellectual nomenclature which we use, and must use, in order to make sense of our individual and collective worlds no longer provides proper guidance. Is your 'soul' in a machine really you? Is it human? Is the human being who evolved in interaction with death human even without death? How to assess the human being when freed of striving and with all needs satisfied? And is the human species still the human species when it evolves away from genetic entropy to homogeneous perfection?

When faced with transcendent issues like these there might be an argument to revert to the common sense perspective, even if common sense has been discredited so many times in human history, particularly on the really big issues. Perhaps common sense is like democracy according to Churchill: the least bad alternative.

[1] Similarly Samuel Scheffler in Death & the Afterlife, 63–64, where he mentions how unaccustomed we are to thinking about humanity's condition in the perspective of billions of years or until the end of the universe.

In fact, common sense might have served better in the past if it had been married to a true recognition of the first principle of epistemology proposed in this book: that reality is almost always wilder and more complex than we imagine.

But, more fundamentally, common sense might be the good guide because common sense invites us to listen to our 'inner voices', and given that obtention of immortality does not allow for a purely reasoned approach the inner voice might be a good guide, the only one available, in fact. David Hume famously said that 'reason is the slave to our emotions, and must always remain so', and although this might be

Reason is the slave to our emotions
(Allan Ramsay: David Hume)

an overstatement, it is probably true that on fundamental metaphysical issues such as earthly immortality all reasoning will be enslaved to the passions – and then one might as well beat a relative straight path to the emotional analysis. Our question is not what the shadows on the wall mean, but what we want them to mean!

When all is said and done our choices on immortality must be driven by the passions, but be informed by the facts, and must heed not only the passion for survival but the whole panoply of passions, for renewal and off-spring, for beauty and the finite, for rest and release!

Chapter 45
Longer Life or Immortality?

The recipe of Epicurus to the quandary of death will appear to many as singularly unconvincing. Death not touching the living and being irrelevant when you are dead seems to be clever sophism. Yet, Epicurus would argue that this is the best we can do. But is that really so?

Is there not an obligation towards life that lessens the pain of the thought and reality of death? An obligation that consists of making life as fulfilling as at all possible, so that death can be encountered with the feeling that you have benefitted all you could – that all your senses have been satisfied. Must our operating principle, much in line with our evolutionary disposition, not be that a good life makes for a good death?

> He who has sought the sweetness of life,
> meets his fate with no fear at all[1]

The human being was designed for life until the thirties. Physical decline sets in earlier, of course, as does mental decline, but even in our thirties most of us are within our design specifications. Our appetite for life is not set to extinguish in the thirties, however, which has good evolutionary reasons and it is surely difficult to satisfy all senses in this timeframe. The mismatch between design life and zest for life has led human beings to gradually shore up durability, to achieve average human life times of forty, then sixty, then seventy and currently over eighty years in some societies. As mentioned, lifetimes have almost doubled within a hundred year time span.

The trouble with appetite is, of course, that it is rarely in sync with need. Obesity constitutes a health crisis in the United States and most of the rich world, and the reason is exactly this mismatch. Appetite for life is no different than the physical sort, it might be suggested, even if we did not truly reach the obesity level with a doubling of average life times within a hundred years. Still, the high incidence of dementia and Alzheimer's at life's end shows that physical possibility and mental

[1] The operetta Farinelli, Den som har livets mildhed soegt.

P. Hulsroj, *What If We Don't Die?*, Springer Praxis Books,
DOI 10.1007/978-3-319-19093-8_45

enjoyment of physical possibility can well part ways. If life time becomes elective surely most human beings will choose a duration of life well beyond 'need', well beyond actual enjoyment. Immortality might not be the up front choice, but how will anybody dare choose death close to the time when the natural arc of enjoyment has been completed? This is the fundamental dilemma of having life on tap, and perhaps even immortality as a choice. Human beings cannot be trusted to choose wisely. Libertarians will argue that 'wisely' is in the eye of the beholder, and that it is none of society's business to render judgment. This may sound rational, but is not, and the fundamental question is in any event not one for rationality, as argued above. However, a 'Big Brother' society where some societal authority will decide whether you have had enough life or not is both unsustainable and morally repugnant. So if Prometheus, each of us, gets the fire he will have to manage it; it cannot be extinguished.

The fundamental issue remains whether Prometheus should seek fire, or should seek only a warmth that will comfort rather than discomfit. Should humans seek immortality or should we seek to extend life expectation to such an extent that every last drop of positive human experience can be squeezed out of life, but nothing more? If the latter is the preferred option, of course, the question becomes what the measure should be for when enough is enough. The answer to this might lie in how aggressively we seek to preserve life. In this sense the Prometheus analogy might be inaccurate, because Prometheus either had fire or not. Yet, if it is agreed that immortality should not be the aim, humankind faces a difficult evolving evaluation process, rather than just a binary choice. The challenge for society would be to push the boundaries of longevity step-by-step, whilst assessing whether additional life continues to equate to additional benefit. And, crucially, to stop the pushing of the boundary when it becomes clear that delta benefit no longer accrues. How this last step can be taken in a democratic process is not so obvious. Still, it might not be impossible, because the generation that will decide to stop the push will normally not be directly affected. Their potential life times would be decided by an earlier generation of medicine and technology, and their decisions would affect progeny, not themselves. To deny your own progeny the possibility of even longer lives is psychologically not self-evident, however, and therefore the perennial question remains whether the genie of longer and longer lifetimes can, in fact, be put back in the bottle once it has been set free. And how firm would any decision be not to pursue immortality, when progeny would sense its salvation if the decision was overturned?

The Bible with its parable of the apple of knowledge would seem to suggest that once the apple is bitten the quest for knowledge is unstoppable. Adam and Eve were ejected from Paradise as a result of their actions, without any indication that the long road of knowledge would ever lead back to Paradise. Yet, in the final analysis the Bible could intimate that the quest for knowledge is a cycle that will ultimately lead back to ignorance: "Blessed are the poor in spirit, for theirs is the Kingdom of Heaven". Given that the meek shall inherit the Earth perhaps the route of Adam and Eve is then one where the quest for knowledge might lead to eternal life on Earth, provided we remain meek in this quest, perhaps even becoming poor in spirit. In the stark perspective of earthly immortality this could be hopeful in a paradoxical manner. Perhaps humanity needs to come to terms with knowledge and reset its destiny to some simpler state, where knowledge is not the dominating feature of existence. Adam might have to spit out a bit of the apple of knowledge he so daringly indulged in!

Disregarding the Bible, the scepticism that the genie can be contained could lead to the conclusion that humankind will pursue immortality, but will be unable to hold on to it once it appears within reach. As alluded to above, the end of the story on immortality could well be that we could resolve the riddle of aging, but not nature's opposition to the ultimate unnaturalness and unhappiness of eternal life on Earth! Finding the happiness gene and twisting it to provide happiness where none can be found might exceed even human ingenuity!

Pursuing the defeatist route even further, perhaps the living conditions of an eternal life should be looked at again. In a future world, where every human need will be taken care of by automated and digitalised means, human effort will become counterproductive. Machines and computers will be able to do everything better. This means that the eternal life of humans will be set against a system where human effort will be welcome only in hobby related areas. Roses might be better cut by non-human means and yet humans might be allowed to tend to roses as a chosen hobby. Football will be played so much better by robots and still humans might indulge in their incapacity as a time-filler. Love will become even more important, but with the reservation that love is a brush that paints on a canvas of life, and with a life of little activity love will be in distress.

In such a world much time and much emotional satisfaction might be sought in the virtual world, well-aided by psychopharfarmacology. The existential question will be whether an eternal life in virtuality will not eventually remove the human being from its moorings. An eternal life of dream might turn into a self-destructive nightmare, where virtual rapture will be the prize. Death-wish might become Selbstaufgabe! The question of how we die moves to the fore!

Chapter 46
Birth Undone

Our lack of concern about not having existed before we were born, and our great concern about not living on after death may lead to reflections on how we would perceive life, and particularly death, if we would live life in reverse. Kierkegaard talked about how life must be lived forward, but can only be understood backwards. But what if our existence started with the horror of death, continued with having the experiences of a whole life at our disposal from the start, but with us seeing every day do away with one day of life until we arrive at the joyful/painful birth, at which time we would return to non-existence, not under the impression of the horror of death, but with the joyful expectation of the life already lived, already celebrated. If our lives started with the horror of extinction and ended with innocence would we value life's experiences more and would our ever-increasing innocence make us move to Wordworth's original state with ease?

Idle speculation, you might say, yet that is not entirely so. In some sense our longer life times have enabled more and more deaths in innocence, the result of the ravages of dementia and Alzheimer's. And our lives have also become far more predictable with advances in medicine and our almost frightening mastery of nature, so that, in fact, humans can plot the broad structure of their lives earlier and much more confidently than in the past.

Ironically, in a twisted sort of way, neither life's greater predictability nor the increased frequency of death in innocence has allowed us to embrace life's arc more readily or appreciatively. The predictability often makes for complacency or disinterest, and the radical emptying of the vessels of life expressed through age-induced dementia makes us panic, not only because of the prospects of the undignified death, but also because we understand this to be a stepped up way of gradually losing ourselves and our existence.

Despite life's boundary conditions having changed, and the changes objectively being in a positive direction, it remains true that the fear of death animates life, and that for most humans the fear of death is inconsolable.

P. Hulsroj, *What If We Don't Die?*, Springer Praxis Books,
DOI 10.1007/978-3-319-19093-8_46

Chapter 47
The Medicated Good Death

If it is accepted that the desire for immortality is often motivated more by fear of the death experience and of non-existence than the wish to live forever, the answer to the question of whether immortality is desirable will depend to a large extent on whether progress can bring about a nicer way to die. A nicer way to die will take some of the sting out of the death experience, and will, possibly, even make the thought of non-existence more bearable.

The possibility of a nicer way to die, in turn, puts centre stage the issue of what the path of life should be and what role we want death to play. If death is the threshold to happy oblivion, but a threshold we are conditioned to resist with every element of our being, perhaps real human progress is not to achieve immortality but to make the crossing of the threshold bearable. This might appear ethically suspect almost in the same way as experimentation with humans, yet is hardly so if the happy crossing becomes a choice rather than an imposition. Searching for the good death might become the ultimate affirmation of life rather than an ugly alchemy striving for the power of darkness.

Prozac and similar medications have put the good death at our disposal, as morphine and opium to some extent did in the past. And yet, we do not prescribe these drugs as a matter of course to those for whom death is imminent. In fact, we tend not even to give them the choice of a relaxed, reality suppressing final experience. The reason for this might be found in our personal morality and is as such objectionable, particularly because we tend to judge in the comfort of a safe distance from our own deaths. If physical courage tends to be emotional ignorance of what the ultimate consequence of the courageous acts might be, why do we deny others a similar kind of emotional ignorance of death's reality, just because the ignorance is medically induced? Our prudish approach to medical happiness in death is surely to some extent motivated by our general rejection of the use of strong drugs to achieve seeming happiness. The battle society is waging against cocaine, LSD, Ecstasy is carried forward to this rather different arena. But surely there is also an existentialist reasoning, no matter how much a painless death in sleep might be desired by most. In a sub-textual fashion we seem to almost agree with von Braun that death should

P. Hulsroj, *What If We Don't Die?*, Springer Praxis Books,
DOI 10.1007/978-3-319-19093-8_47

Tod und Verklärung
(Rafael: Transfiguration)

be lived and observed. We believe that death is about Tod und Verklärung, as Richard Strauss's tone poem so beautifully puts it. We believe ultimately in not disturbing nature's arc of birth to death!

In an intellectually very uneven book,[1] Hans Küng has analysed near death and after death experiences from the religious perspective. If one disregards the religious perspective one would appear to get lots of confirmation of why a good death may not be a medicated death. Verklärung is writ large, although Küng accepts that this Verklärung might ultimately be physically induced. Hence we come full circle. If Verklärung is physically induced, why do we, fellow citizens, believe that it is for us to judge whether a Prozac induced end is desirable or not? Should not every dying person be given a choice, unbiased by our own morality, of the kind of death this human being wishes to die?

When looking at the practical implications of earthly immortality we find problems and much to be disliked. Looking at ever-longer life times we find much to like, as long as we do not allow ourselves to over-indulge. Looking at the good death we find that progress has given us new options, and will give us more. If the remedy against over-indulgence is a good death, society should be intent on providing as many options as possible, and allow well-informed geriatrics to choose their own best end! Immortality might not be the Holy Grail. A long life and a good death might!

[1] Eternal life? Life after death as a medical, philosophical, and theological problem.

Chapter 48
Death Elective

The conundrum of immortality may be resolved if life-time and hence death become elective, and if death loses most of its sting. Society might pursue its relentless drive for pushing the boundary farther and farther, as long as it happens hand in hand with making death less and less frightening. Perhaps such a model could become the epitome of individualistic responsibility; a libertarian dream come true: the human being choosing its life-time and choosing its death and its way of dying. Big Brother enables but does not decide.

Some humans will choose shorter life times, some longer, some perhaps even eternal life. The result would be the ultimate liberation from the dictates of life and death, and the ultimate empowerment of the individual. Perhaps in itself a completion of an arc that started with individualistic anarchy at the dawn of human time, over ever increasing collectivity in order to fight the dictates of nature and the challenges of existence, to the freedom to choose life and to choose death, largely removed from the tyranny of nature or others. Wish it was that easy!

The freedom that will be gained by death elective is of the most terrifying kind: choosing rightly might win you everything, choosing wrongly might make you lose impossibly much. It poses the ultimate question of when the choice of suicide is right, and thus puts reality to the dilemma treated by Albert Camus in The Myth of Sisyphus of whether choosing death is right. Camus answers no, although it is to be doubted that this answer truly considers the life situation of Sisyphus, who must eternally push the stone up the mountain, only to see it roll down again. Using this as an allegory for the human condition, Camus rejects suicide, but this is, in the final analysis, informed by the knowledge that the human condition involves ultimate death. Camus could hardly advise Sisyphus to continue his effort forever, if Sisyphus could have escaped through death after having had more than his fill of pushing. Camus states that one must imagine Sisyphus happy – yet it must be assumed that Sisyphus would have been even happier dead, had he had that option!

P. Hulsroj, *What If We Don't Die?*, Springer Praxis Books,
DOI 10.1007/978-3-319-19093-8_48

'One must imagine Sisyphus happy'… Forever?
(Tizian: Sisyphos)

Death elective would put us in the existentialist angst situation of Kierkegaard and Sartre; it would give us the burden of continually having to decide for life or for death; it would leave us not only to deal with the fear of death (albeit perhaps medicated), but also with the pull of death, as the person at the cliff's edge, torn between the fear of falling and its strange fascination. Camus would perhaps have argued that this is the human condition already now, yet the difference is that when death becomes elective, all deaths will be suicide. With Paul one could then, in a very round-about way, say: Death is swallowed up in victory. O death, where is thy sting? O grave, where is thy victory?

Janis Joplin sang that 'freedom's just another word for nothing left to lose'. Elective death is the opposite: you have it all, including the choice to leave it all, and that sort of ultimate empowerment might be properly understood as freedom, or might be understood as a tyranny of choice, not just once but all the time until the choice is suicide. Every time a person chooses life the person will also have chosen to have to make yet another choice – repeated until the choice of death. Not so different in the end from Sisyphus's pushing of the stone, and, from a Camus perspective, making existence even more absurd. Even so, elective death is not necessarily life denying, but pushes us, again, to give lie to rationality and reason, and invites us to find joyful meaning in an increasingly meaningless existence. Thus, ironically, the rational choice for death is when that struggle against the meaninglessness is lost by an individual, because sensory joys, be they immediate in the Wordsworth, Ode to Immortality sense, or reflective and intellectual, hold no attraction any longer. Death of joy leading to death of being.

The premise of liberal democracy is that freedom, exercised collectively or individually, is the best recipe for happiness and good outcomes. This is not necessarily so! Economic and cultural individualism has often led individuals to disaster, and democracy often led to bad outcomes. Democracy might be the least bad system and liberalism might overall be better than all alternatives, yet the burden of choice can be heavy. But paradoxically, the availability of choice in combination with prosperity can also lead to indifference as the current state of democracy so amply shows in the West, and as the lack of resolve in so many rich kids also demonstrates. The eternal alternative constituted by elective death might therefore also lead to heavy burdens and plenty of bad decision making, and, worryingly in the light of possible eternal unhappiness, to the default decision of indecision. Humankind has always been torn between the creative power of choice and the comfort of authoritarianism, the latter making public indecision an art form. However, authoritarianism relative to the timing of individual death is frightening, prone to abuse and unsustainable, and hence when we take the route of possible immortality our only option is to take with it the ultra-individualism of having to choose one's own death!

The attractiveness of the eternal alternative is, however, not only the self-determination and the outcomes it leads to but also the uplifting nature of eternal choice. Decisions will be continually required and every time a person chooses life it will be an affirmation of the joy of life. And, to a limited extent, that is true even when the choice is only the choice of indecision. Existentialism becomes a life enhancer in the vein of Kierkegaard, not only a coping mechanism à la Camus.

This sort of existentialism is not necessarily premised on a belief in free will, it is premised on having a substantive choice, which we might be able to exercise freely or not. No matter how freely, the exercise of the choice becomes the life-enhancer. Elective death is a fork in the road and will appear so to a driver, even if the driver for his or her own reasons turns out to be able only to steer right!

Chapter 49
Blessed Are Those Who Have Not Seen and Yet Have Believed

If progress brings immortality within reach one may wonder if progress will not also make us able to establish whether god exists or not. Also here the first principle of epistemology might rule, even if one can speculate that an all-powerful god would always make Icarus crash. Still, the need for the great leap of faith in order to attain divine atemporality might disappear if science would show humans the face of God. Perhaps the path of faith is not just to believe without having seen, but a migration from paradise to paradise, having been expelled by biting the apple of knowledge and finding the way back through that same bite.

There might be a very depressing end to the journey of knowledge, humankind returning to original matter, or there might be a climatically joyful one, where progress will not only let us defeat death, but will let us become even more god-like than just having been 'made in His image'; progress letting us rejoin God without passing the threshold of death blind to what to expect beyond. Current talk of the god gene is an unsuccessful attempt to use science to resolve the god-question, but who knows where a multitude of such lines of inquiry will take us? Plotted it cannot be, but also it cannot be excluded that the existence of God, perhaps through the will of God, will be divulged. What, after all, was the point of letting Thomas put his hand in the wounds? Humankind might for thousands of years have been limited to believe without having seen, but rapture might not be as spontaneous as believed. Science might eventually bring an entirely different kind of rapture, and allow the growing legions of doubting Thomases to finally see and fell the splendour of God! And similarly science might be able to prove or disprove re-birth and Nirvana, might be able to generally move the meta-physical to the physical! Kant might have to rethink his 'Kritik der reine Vernunft'.

P. Hulsroj, *What If We Don't Die?*, Springer Praxis Books,
DOI 10.1007/978-3-319-19093-8_49

Chapter 50
So …

It goes with the territory of the proposed first principle of epistemology that the human quest for immortality could take completely different directions than those laid out in this book. Yet, it also goes with the territory that future reality is likely to be much wilder and more complex than foreseen here. It could be argued that being true to the principle one should not try to divine the indivinable. And there is some truth to that. Humankind should never put all its eggs in the basket, for instance, of believing that ultimately we will see the face of god here on earth, and this irrespective of whether god exists or not. If he exists he might not allow us to see him. A literal reading of the Bible could lead to that conclusion.

But it is ultimately perhaps also unsound to believe too strongly in humans reaching the Holy Grail of immortality, and to draw conclusions from that too firmly.

What is sound, however, is to reflect on what we do not want of the future that we ourselves will create. A sound methodology is to abstain from too many positive determinations that might prove utterly wrong, and instead concentrate on the negative determinations that it will be within our gift to avoid.

High on the list of negations should be machine-based life, avatars with our minds, networked consciousness, and purely virtual lives. If we allow the defining features of what it means to be human to become too blurred we will lose humanness altogether, and why would we want that, when the quest for immortality is exactly to preserve to the utmost our humanness? Evolution romanticism might militate against this view, since evolution has led to so many species becoming extinct and replaced by better performing species, yet why a species would seek its own destruction when to a large extent it can call the shots would be hard to explain even from the perspective of evolution romanticism. A species paradigm-shift might make sense for those who would believe that this would save some part of themselves, but makes no sense from a gene survival or a broader species perspective. Hence, for humankind this route should not be navigated. Humans should remain human.

P. Hulsroj, *What If We Don't Die?*, Springer Praxis Books,
DOI 10.1007/978-3-319-19093-8_50

In the same vein we should not wish for longer and longer, or eternal, life, if such expanded lives were not filled with human content as we know it now. An eternity of boredom, repetition and spiritual death we should not seek; an eternity of love and sensory and intellectual discovery would be wonderful, but perhaps only obtainable if humans can reach atemporal immortality. But who knows, the first principle of epistemology might yet again surprise us.

In the final analysis the quest for immortality should perhaps instead be understood as a quest for evermore joy in its truest sense, for evermore spiritual satisfaction. And hence, when joy and spiritual satisfaction subside, so should life! Life should end then, but even the end should be similarly informed. The quest for spiritual satisfaction, if not joy, will have to be pursued for death as well as for life!

Chapter 51
An Afterthought – The Fifth Dimension

Immortal life on earth must be based on sempiternity. It is hard to imagine that humans, on their own, can undo the texture of time and provide for earthly atemporality. And even if they could, what would it help? We would then just be faced with the dilemmas that make it so hard to understand if immortality in the heavens is really desirable.

It is wrong, however, to discount the possibility or desirability of heavenly immortality too quickly. With the human analytical apparatus we can see only two possibilities of immortality in the beyond, temporal or atemporal. Doing so, we possibly ignore the shackles of our conceptual ability, we possibly ignore that the reality we perceive is conditioned by evolution, and that this conditioning might have hard-coded an inability for us to grasp other typologies of reality, because a broader understanding of reality could be detrimental to gene survival.

Apparently many people, including the cardinal referred to in an earlier chapter, believe that eternal life in the beyond is just a continuation in more pleasant circumstances: viz. 'John-Paul is now talking to God for the first time'. This is not hard to understand, but hard to believe. Yet, already with the idea of atemporality our ability to understand is severely challenged. Therefore, perhaps intuitively recognising the shackles of imperfect comprehension, some of us have never really tried to give substance to immortality, but have left it as an abstract notion of some sort of something; possibly detecting signs of immortality but recognising that it goes beyond human capability to understand.

Leaving the notion of immortality unexplored might thus be an instinctive recognition that immortality might be unexplorable, might be a recognition that if immortality exists then it is, and will always remain a fifth dimension for human beings, which we are not equipped to understand. The first principle of epistemology of this book invites human humility, and in relation to the concept of divine immortality perhaps the principle dictates the singularity; intimates that we will never be able to look beyond the black hole of our inbred sense of dimensions. Immortality might be a divine concept shrouded in a veil that will be forever unpierceable for humankind.[1]

[1] Somewhat in this direction Augustine in the Confessions.

P. Hulsroj, *What If We Don't Die?*, Springer Praxis Books,
DOI 10.1007/978-3-319-19093-8_51

Immortality in this sense will, in any event, never be re-creatable on Earth, unless humans with time gain the ability to move not only between different universes but into perhaps untold numbers of dimensions. And that is a tall order, indeed.

Immortality as a fifth dimension?
(Kazimir Malevich, Black Square)

However, immortality as a fifth dimension of the heavens responds beautifully to the feeling that the seed of immortality has been sown in us. The seed might, of course, be a smokescreen only, designed to avoid us peering consciously into the abyss of perdition. Or it might be a seed from which we are not able to divine the shape of the eventual tree, simply because there may be things that humans are not meant to understand: the existence of God or the reality and nature of immortality. Perhaps death is the door to other realities. Perhaps death removes the blinkers evolution has put on us in term of understanding all reality, not just the reality that serves our earthly survival. Hindu and Buddhist thought may be understood as explaining a constant fight against the limitations put by evolution, with each rebirth shackling the individual once again until ultimate release into Nirvana, which then perhaps can be understood as a participation in all dimensions of reality!

Not only from the perspective of Hinduism and Buddhism but also from the perspective of evolution, there is a possible tragedy then in vying for earthly immortality because it would perpetuate our inability to embrace all of reality.

What is certain is that the embrace of immortality as a fifth dimension requires a leap of faith, a leap of faith even greater than the one of Soeren Kierkegaard!

Printed in the United States
By Bookmasters